blueprint reading for construction

Residential and Commercial

write-in text

by

WALTER C. BROWN

Professor Emeritus,
Division of Technology
Arizona State University, Tempe

Publisher

THE GOODHEART-WILLCOX COMPANY, INC.

Tinley Park, Illinois

STEPS IN READING A SET OF BLUEPRINTS

Step 1 — Check the list of blueprints in the set.

Step 2 — Study the plot plan to observe the location of the building.

Step 3 — Check the first floor plan to orient the building.

Step 4 — Observe features such as the entry and hallways, relation of rooms, windows, doors and special features like offsets and raised ceilings.

Step 5 — Study features that extend for more than one floor, such as plumbing, stairways and fireplaces.

Step 6 — Observe floor and wall construction and other framing details.

Step 7 — Check the foundation plan and the detail sheets.

Step 8 — Study the mechanical blueprints for details of plumbing, air conditioning and electrical.

Step 9 — Check the notes and specifications against construction details.

Step 10 — Move from one sheet to another until you are thoroughly familiar with the structure.

Library of Congress Catalog Card Number 90-42584
International Standard Book Number 0-87006-825-3

6 7 8 9 10   90   96

Library of Congress Cataloging in Publication Data

Brown, Walter Charles.
    Blueprint reading for construction: residential and commercial: write-in text/Walter C. Brown.
        p.   cm.
    Includes index.
    ISBN 0-87006-825-3
    1. Building—Details—Drawings. 2. Blueprints.
I. Title
TH431.B76   1990
692'.1—dc20   90-42584
                                   CIP

INTRODUCTION

BLUEPRINT READING FOR CONSTRUCTION is a training course for those who desire a knowledge of basic blueprint reading or increased knowledge of construction drawings. The term "Blueprint Reading" as used in this write-in text, refers to interpreting and visualizing construction drawings whether or not the drawings actually are blueprints.

BLUEPRINT READING FOR CONSTRUCTION is a combination text and workbook. The text tells and shows HOW, and the workbook provides space for meaningful blueprint READING, SKETCHING and ESTIMATING ACTIVITIES. Actual construction blueprints used with the text can be found in the Large Prints Folder. They provide the student with realistic job experiences. The text is equally applicable for students in actual construction, estimating or construction management. All must be able to read construction blueprints.

BLUEPRINT READING FOR CONSTRUCTION is intended for technical students, apprentices and adult workers. It also is designed to be used for self-instruction by individuals unable to attend classes.

Walter C. Brown

CONTENTS

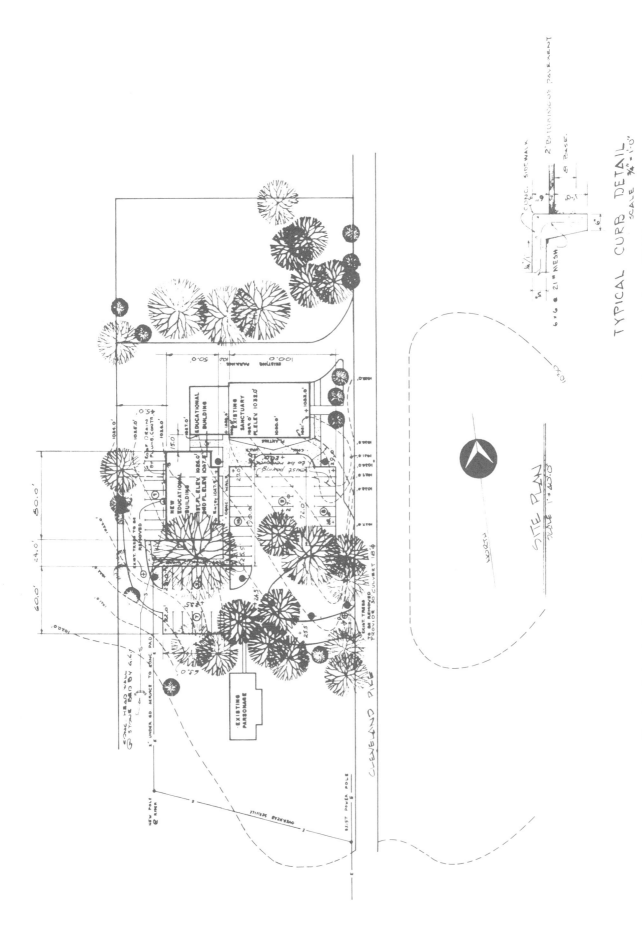

Fig. 1-1. A site plan showing roof outline and location of principal features on the lot. (FHA)

PART 1
INTRODUCTION TO CONSTRUCTION BLUEPRINT READING

Unit 1
Importance and Use of Construction Blueprints

Construction blueprints are a means for the owner, architect and builder to communicate ideas and relate specific directions. Blueprints, together with the set of specification sheets, detail what is to be built, what materials are to be used and how the job will be done.

The set of blueprints or working drawings form the basis of agreement and understanding that a building will be built as it is planned. Therefore, all persons who are involved in the planning, supplying and/or building of any structure should be able to read construction blueprints.

This text will help you learn to read blueprints. In addition, it is designed to aid you in expressing construction ideas and details by means of freehand sketches. This unit tells what a construction blueprint is; what comprises a set of prints; how prints are made and how to care for them.

Construction Blueprints

A BLUEPRINT is a COPY of a DRAWING which tells the owner, architect and builder what the structure will look like when it is completed. For many years, the blueprint was the only type of reproduction used. It consisted of a print with white lines on a blue background. That is, the print was primarily blue, so it was called "blueprint." Today, the term "blueprint" is widely used to refer to all types of copies of construction drawings. The prints have changed to dark lines on a light background. See Fig. 1-1.

Blueprints, sometimes referred to as a set of WORKING DRAWINGS, provide the details of size and shape description, materials to be used, finish and other special details of construction. Occasionally, blueprints are referred to as PLANS, but this term has a more restricted meaning. In the more restricted sense, "plan" refers to those views taken from directly above the object such as "floor plan" or "site plan."

A set of blueprints usually consists of the following sheets:

The SITE PLAN shows the location of the building on the site, Fig. 1-1. The plan also may include topographic features such as contour lines and trees, and construction features such as walks, driveways and utilities. Frequently, the roof plan for the structure is shown on the site plan.

The FOUNDATION AND BASEMENT PLANS usually are included on the same plan when a basement is a part of the structure. See Fig. 1-2. This plan includes the foundation walls, footings, piers, fireplaces, stairways, partition walls and other installations such as bath fixtures or other built-ins. Details and section drawings of the foundation wall and footings are sometimes included on the same sheet.

The FLOOR PLAN, perhaps, is the most important drawing of all in that it provides the largest amount of information. Study Fig. 1-3. The floor plan actually is a sectional view taken on a

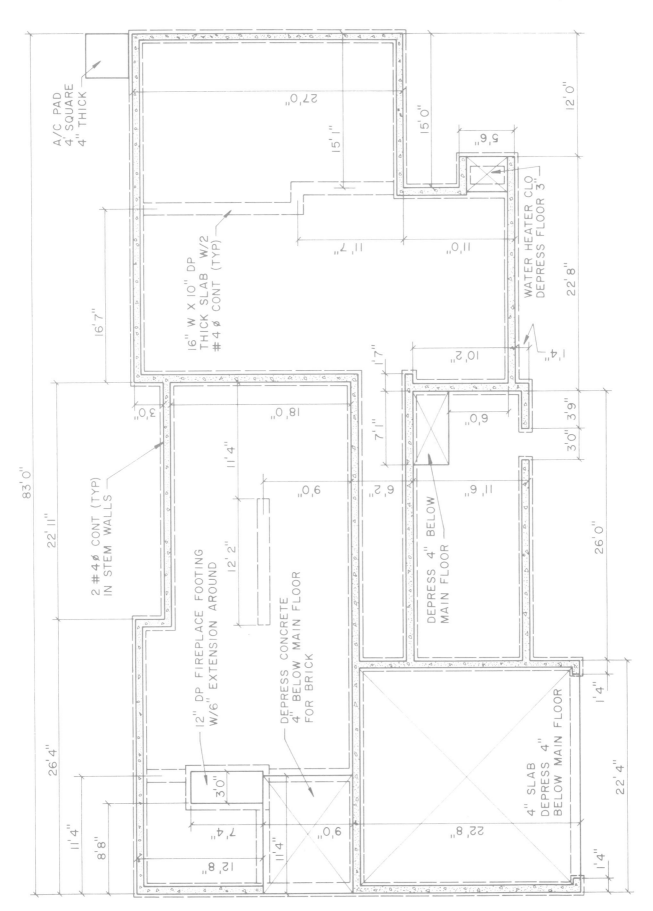

Fig. 1-2. A foundation and basement plan includes footings, columns, foundation walls and other details.

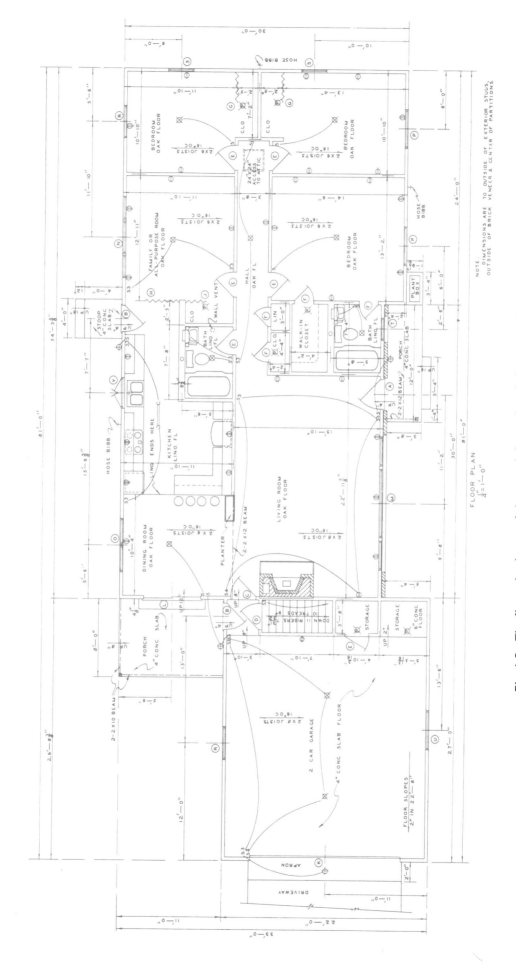

Fig. 1-3. The floor plan is one of the most informative of all construction drawings.

9

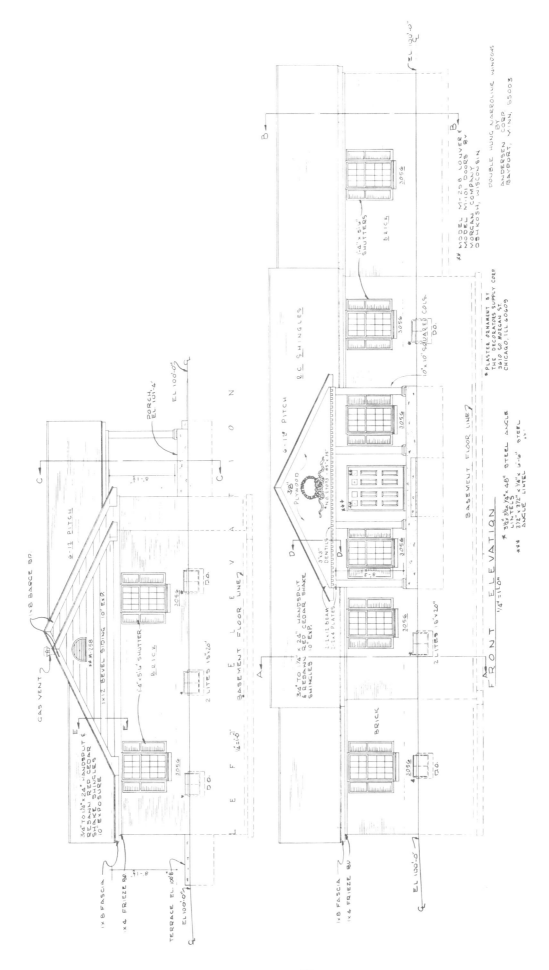

Fig. 1-4. This elevation drawing presents the building from two of several view points.

horizontal plane midway between the floor and ceiling. It shows exterior and interior walls, doors, stairways, fireplaces, built-in cabinets and mechanical equipment. Drawings for multistory buildings would include a floor plan for each floor.

ELEVATION drawings are the views of the building showing its exterior features. See Fig. 1-4. Usually, four elevation drawings are needed to show the design of all sides. More elevation views are required for buildings of unusual design, such as those with more than four sides or internal courtyards. Elevation drawings would be drawn of interior view, but these usually are referred to as "detail" drawings.

SECTIONS are drawings showing the construction of walls, stairs or other details not clearly shown on the elevation, floor plan or framing drawings, Fig. 1-5. These sectional views are known as detail sections and usually are drawn to a scale larger than that used for the elevation and floor plan drawings. Sectional views taken through the narrow width of an entire building are known as "transverse" sections. Those through the long dimension are known as "longitudinal" sections.

DETAIL DRAWINGS usually are required where special or unusual construction is to be performed.

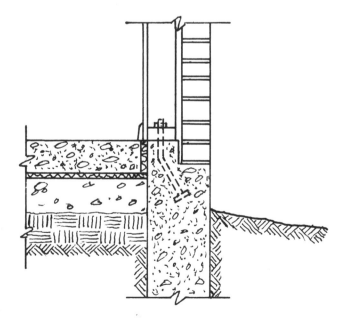

Fig. 1-5. A typical sectional view illustrates the construction of a ground-supported slab floor.

Frequently, these are detail views of some special feature such as an arch, cornice, doors, windows or retaining wall drawn to a larger scale to clearly describe the manner in which the construction is to be performed. See Fig. 1-6.

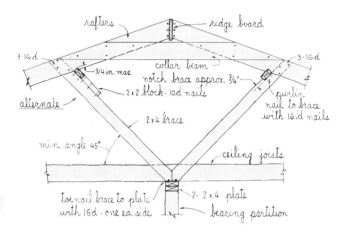

Fig. 1-6. Detail drawing shows special framing for rafter bracing. (FHA)

DOOR AND WINDOW SCHEDULES are not drawings, but usually are included as a part of a set of working drawings, Fig. 1-7. These schedules consist of a ruled form or enclosure with columns for listing the necessary information and sizes for specifying the various type doors and windows included in the construction. Each item in the schedule is referenced to the plan and elevation drawings. Door schedules frequently are included on the plan drawings. Window schedules generally appear on the elevation drawings.

The ELECTRICAL PLAN may appear on the floor plan itself for normal construction jobs. See Fig. 1-3. For more complex structures, a separate plan, omitting unnecessary details and showing the electrical layout, is added to the set of plans. This is called the electrical plan. Location of the meter, distribution panel, switches, convenience outlets and special outlets, (such as door chimes, telephones, etc.) are shown on the electrical plan.

The AIR CONDITIONING PLAN, like the electrical plan, is included on the regular floor plan for simple air conditioning (heating and cooling) installations. For more complex jobs, a separate A/C plan is added to the set of plans. This is a special floor plan

DOOR SCHEDULE

SYMBOL	QUANTITY	TYPE	DOOR SIZE	REMARKS
A	1	Panel	3' – 0'' x 6' – 8'' x 1 3/4''	Fir
B	1	Flush	2' – 6'' x 6' – 8'' x 1 3/4''	Birch, solid core
C	1	Panel	2' – 6'' x 6' – 8'' x 1 3/4''	Fir w/2 hammered glass lites
D	1	Panel	2' – 6'' x 6' – 8'' x 1 3/4''	Fir w/ventilating lite
E	1	Panel	2' – 6'' x 6' – 8'' x 1 3/4''	Fir
F	2	French	2' – 6'' x 6' – 8'' x 1 3/4''	Fir
G	1	Glass Sliding	9' – 0'' x 6' – 10''	Aluminum
H	2	Flush	2' – 0'' x 6' – 8'' x 1 3/4''	Masonite, w/louvres
J	4	Louvre	1' – 4'' x 6' – 8'' x 1 3/8''	Fir
K	3	Bifold Louvre	5' – 0'' x 6' – 8'' x 1 3/8''	Fir, track at top only
L	4	Panel	2' – 6'' x 6' – 8'' x 1 3/8''	Fir
M	2	Panel	1' – 10'' x 6' – 8'' x 1 3/8''	Fir
N	1	Flush	2' – 6'' x 6' – 8'' x 1 3/8''	Birch, hollow core w/louvre
O	2	Panel	2' – 4'' x 6' – 8'' x 1 3/8''	Fir
P	1	Pocket Louvre	2' – 0'' x 6' – 8'' x 1 3/8''	Fir
R	1	Panel	2' – 0'' x 6' – 8'' x 1 3/8''	Fir
S	1	Panel Dutch	2' – 6'' x 6' – 8'' x 1 3/8''	Fir

WINDOW SCHEDULE

1	2	Sliding	4' – 0'' x 4' – 0''	Aluminum
2	1	Sliding	4' – 0'' x 3' – 0''	Aluminum
3	1	Sliding	2' – 0'' x 3' – 0''	Aluminum
4	4	Ventilating	6' – 0'' x 6' – 10''	Aluminum
5	1	Fixed	6' – 0'' x 6' – 10''	Aluminum
6	1	Ventilating	2' – 8'' x 6' – 10''	Aluminum

Fig. 1-7. Door and window schedules provide information for use in specifying types needed for a construction job.

on which more complete air conditioning installation details are shown.

The PLUMBING SYSTEMS PLAN shows the layout for the piping system that supplies the hot and cold water, the sewage disposal system and the location of plumbing fixtures. Plans for small residences may include the entire plumbing system plan on one drawing. Frequently, plumbing fixtures are shown on the floor plan for a residence. For more complex structures, separate plans for each system may be used.

FRAMING PLANS may be included in a set of plans for the framing of the roof, floors and various elevations or wall sections, Fig. 1-8. These plans are required for more complex structures, but may be omitted for smaller, less complicated buildings.

How Prints Are Made

The original drawing, called a TRACING, usually is made on a transparent paper or polyester film. The tracing is placed over a sheet of sensitized paper and exposed to ultraviolet light, Fig. 1-9.

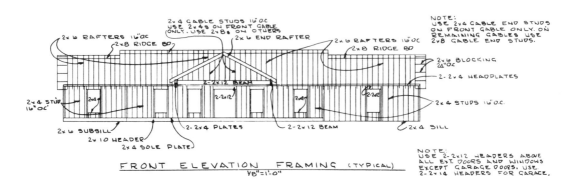

Fig. 1-8. A framing plan of front elevation for a residence.

Fig. 1-9. Copies of original drawings are made on a copy machine. The machine shown here is a whiteprinter—diazo type. (Bruning Division)

Then, the exposed paper is fed through a developing process, and it becomes a PRINT. The type of paper selected (the nature of its light sensitive surface) will determine the type of print produced, a blueprint or a "whiteprint" with blue, black, red or brown lines.

Reading a Blueprint

Blueprint reading is the gathering of information from a blueprint. It involves two principal elements: visualization and interpretation.

Visualization is the ability to "see" or envision the size and shape of the object from a set of blueprints that shows several views of the building. A study of blueprint reading principles and learning to do freehand sketching (see Unit 5) will help you visualize the details of construction from a set of blueprints. Another important aspect of blueprint reading is the ability to interpret lines, symbols, dimensions, notes and other information on the print. Each of these factors will be presented to you in this text in logical order to help you learn how to read the blueprints used in construction today.

Care of Blueprints

Blueprints and related specification sheets are as important as the tools you use. With proper care, blueprints can be kept usable for a long period of time.

Rules of care that you should observe are:

1. Never write on a print unless you have been authorized to make changes.
2. Keep prints clean and free of oil and dirt. Soiled prints are difficult to read and contribute to errors.
3. Fold or roll prints carefully to avoid tearing.
4. Do not lay sharp tools or pointed objects on prints.
5. While in use, lay prints in a safe and secure place to avoid being stepped on or damaged by the wind.
6. When not in use, store prints in a clean, dry place.

Blueprint Reading Activity 1—1
USE OF CONSTRUCTION BLUEPRINTS

1. Study the DICTIONARY OF TERMS and ABBREVIATIONS in the REFERENCE SECTION. Familiarize yourself with these sections and their location for future reference.

2. Look over some of the blueprints contained in this text to gain an idea of the nature of construction blueprints.

3. Be prepared to tell about any experiences you have had in using blueprints at home, school or work.

4. How many different kinds of prints have you seen or used? Describe these.

5. The term "blueprint" originally referred to a certain type of print. Describe this print. What does the term refer to today?

6. What other terms are used to refer to a set of blueprints?

7. Identify and describe the various sheets of prints which may be included in a set of blueprints.

8. Reading blueprints involves two elements. What are they? Explain each.

9. Of what importance is the proper care of blueprints?

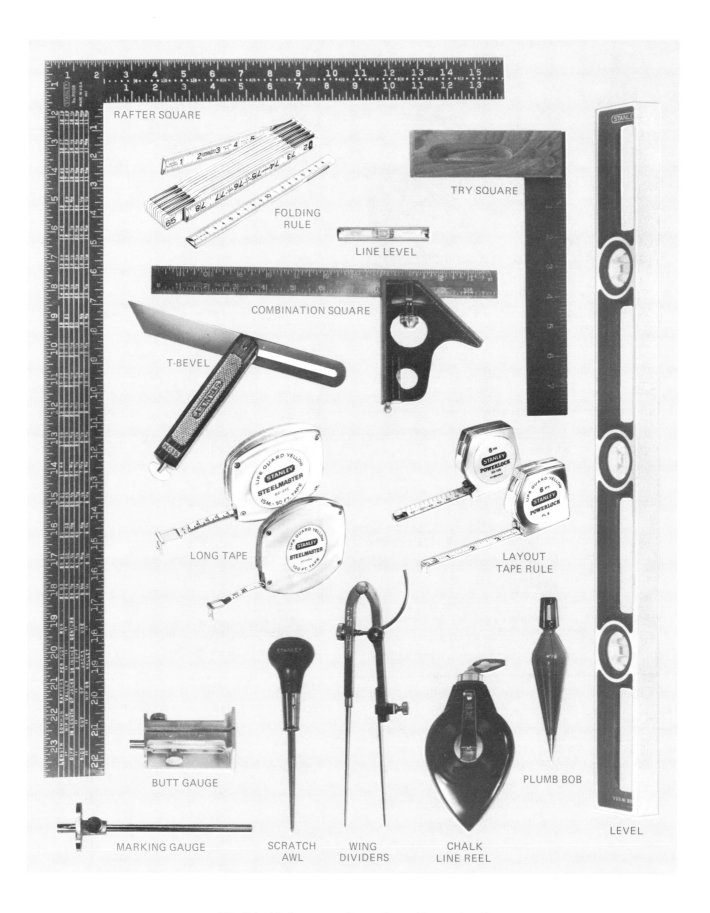

RAFTER SQUARE

FOLDING RULE

TRY SQUARE

LINE LEVEL

COMBINATION SQUARE

T-BEVEL

LONG TAPE

LAYOUT TAPE RULE

BUTT GAUGE

PLUMB BOB

MARKING GAUGE

SCRATCH AWL

WING DIVIDERS

CHALK LINE REEL

LEVEL

Fig. 2-1. Various measuring tools used in construction.
(Stanley Tools)

Unit 2
Measuring Tools in Construction

Measuring tools directly related to reading blueprints and laying out measurements are the framing square, bench rule, steel rules and tapes, Fig. 2-1. This unit is designed to help you review the methods of reading the various measuring tools used in construction.

In the English, or customary, measurement system, the distances are divided into feet, inches and fractional parts of an inch. The rule used with this system is called the "fractional rule." In the metric system, the divisions are in meters, centimeters and millimeters. The rule is called the "metric rule."

Fractional Rule

Measurements in the construction industry are seldom closer than an eighth of an inch (customary system). Therefore, the fractional rule usually is divided into 8ths or 16ths. On the rule shown in Fig. 2-2, the inch is divided into 16 parts, and each small division is 1/16th of an inch in size. To read the fractional rule, start with the edge divided into 16ths and follow these steps:

1. Study the major divisions of the inch numbered 4, 8 and 12. These represent 4/16, 8/16 and on to 16/16 or 1 inch. There are four of these major divisions in an inch; therefore, each one is equal to 4/16 or 1/4 of an inch.
2. Note that there are four small divisions in each major division of 1/4 inch. Each small division represents 1/16 of an inch. Two of these divisions equals 2/16 or 1/8.
3. Further study and application of fractional parts will enable you to locate any common fraction that is a multiple of 16ths. For example:

$$\frac{4}{16} = \frac{1}{4}; \frac{8}{16} = \frac{1}{2} \text{ and } \frac{10}{16} = \frac{5}{8}$$

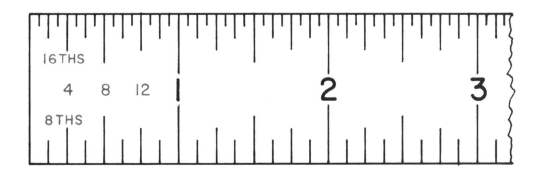

Fig. 2-2. Typical divisions of a fractional rule.

Complete the readings called for in the activities that follow. Reduce to the lowest terms and place your answers in the spaces provided.

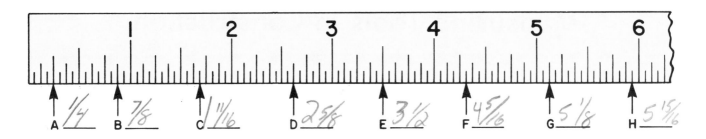

Metric Rule

The unit of linear measure in the metric system is the meter (m). In the building and construction fields, the centimeter (1/100 m) is the official unit of measurement. Building material sizes are given in millimeters or centimeters. For example, the nominal size of the metric brick (including joint thickness) is 225 mm x 112.5 mm x 75 mm. The standard modular sheet size is 120 cm x 240 cm. Metric measurements in cabinet work usually are given in centimeters and, occasionally, in decimal fractions of the centimeter (or millimeters).

The smallest subdivision on the metric framing square is 1/2 centimeter. On tapes, the smallest subdivision usually is the millimeter. To read the metric rule, follow these steps:
1. Study the major divisions marked 1, 2, 3, etc. Each of these represents 1 centimeter (cm), Fig. 2-3.
2. Each centimeter is divided into 10 millimeters.
3. A meter is 100 centimeters, which is marked 1 m on the rule.
4. For a reading of 3.4 meters, follow the rule to three meters (3 m) and on to 40 cm (.4 m), Fig. 2-3.

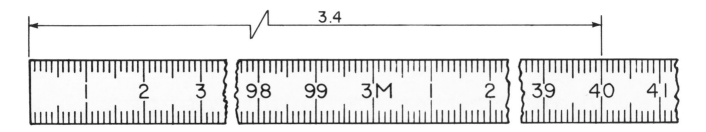

Fig. 2-3. A measurement of 3.4 meters is shown on the metric rule.

Measurement Activity 2−2
READING THE METRIC RULE

Complete the readings called for in this activity. Place your answers in the spaces provided.

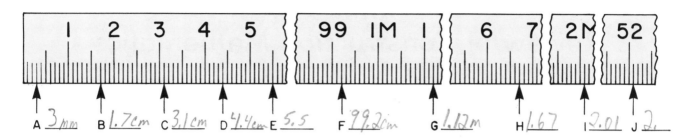

A ___3mm___ B ___1.7cm___ C ___3.1cm___ D ___4.4cm___ E ___5.5___ F ___99.2cm___ G ___1.12m___ H ___1.67___ I ___2.01___ J ___2.___

Unit 3
Review of Construction Mathematics

Construction workers and estimators frequently need to make calculations in connection with the reading of blueprints. This unit deals with construction-oriented calculations involving common fractions and decimals in customary measurements and decimals in the metric system.

Common Fractions

Common fractions are written with one number over the other, such as $\frac{11}{16}$. The number on the bottom, 16, is called the DENOMINATOR. It indicates the number of equal parts into which a unit is divided. The number on top, 11, is called the NUMERATOR. It indicates the number of equal parts taken, Fig. 3-1. In the fraction shown, $\frac{11}{16}$, eleven of the sixteen parts are taken.

A PROPER FRACTION is one whose numerator is less than its denominator, as: $\frac{7}{16}$ and $\frac{3}{4}$.

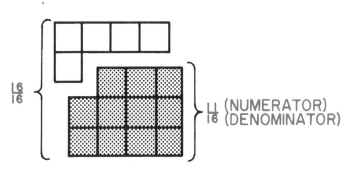

Fig. 3-1. Unit divided into 16 fractional parts.

An IMPROPER FRACTION is one whose numerator is greater than its denominator, as: $\frac{5}{4}$ and $\frac{19}{16}$.

A MIXED NUMBER is a number which consists of a whole number and a proper fraction, as: $2\frac{3}{4}$ and $5\frac{1}{8}$

Fundamental Steps in the Use of Common Fractions

1. Whole numbers may be changed to fractions by multiplying the numerator and denominator by the same number:

   Change 6 (whole number) into fourths.

   $$\frac{6}{1} \times \frac{4}{4} = \frac{24}{4}$$

   Each whole unit contains 4 fourths.

   Six units will contain 6 x 4 fourths or 24 fourths.

   The value of the number has not changed,

   $$\frac{24}{4} = 6.$$

2. Mixed numbers may be changed to fractions by changing the whole number to a fraction with the same denominator as the fractional part of the mixed number and adding the two fractions:

$$3 \frac{5}{8} = \left(\frac{3}{1} \times \frac{8}{8}\right) + \frac{5}{8}$$

$$= \frac{24}{8} + \frac{5}{8}$$

$$= \frac{29}{8}$$

Each whole unit contains 8 eighths.

Three units contain 3 x 8 eighths or 24 eighths.

Adding the $\frac{5}{8}$ part of the mixed number to $\frac{24}{8}$ gives us $\frac{29}{8}$.

3. Improper fractions may be reduced to a whole or mixed number by dividing the numerator by the denominator:

$$\frac{17}{4} = 17 \div 4 = 4 \frac{1}{4}$$

4. Fractions may be reduced to the lowest form by dividing the numerator and denominator by the same number:

$$\frac{6 \div 2}{8 \div 2} = \frac{3}{4}$$

The value of a fraction is not changed if the numerator and denominator are divided by the same number, since this is the same as dividing by 1.

5. Fractions may be changed to a higher form by multiplying the numerator and denominator by the same number:

$$\frac{5 \times 2}{8 \times 2} = \frac{10}{16}$$

The value of a fraction is not changed by multiplying the numerator and denominator by the same number, since this is the same as multiplying by 1.

Addition of Fractions

To add common fractions, the denominators must all be the same:

Example: $\frac{5}{16} + \frac{3}{8} + \frac{11}{32} = ?$

The LOWEST COMMON DENOMINATOR into which these denominators can be divided is 32.

Change fractions to a higher form (Fundamental Step No. 5)

$$\frac{5}{16} \times \frac{2}{2} = \frac{10}{32}$$

$$\frac{3}{8} \times \frac{4}{4} = \frac{12}{32}$$

Add the fractions with the common denominators

$$\frac{10}{32} + \frac{12}{32} + \frac{11}{32} = \frac{33}{32}$$

Reduce improper fraction (Fundamental Step No. 3)

$$\frac{33}{32} = 1 \frac{1}{32}$$

Construction Mathematics Activity 3—1
ADDITION OF FRACTIONS

Solve the following problems. Reduce answers to lowest form.

1. $\frac{3}{4} + \frac{1}{8} + \frac{1}{2} =$ $\frac{6}{8} + \frac{1}{8} + \frac{4}{8} = \frac{11}{8} = \left(1\frac{3}{8}\right)$

2. $\frac{7}{8} + \frac{3}{16} =$ $\frac{14}{16} + \frac{3}{16} = \frac{17}{16} = \left(1\frac{1}{16}\right)$

3. $\frac{5}{12} + \frac{3}{8} + \frac{3}{4} =$ $\frac{10}{24} + \frac{9}{24} + \frac{18}{24} = \frac{37}{24} = \left(1\frac{13}{24}\right)$

4. $\frac{3}{10} + \frac{9}{10} + \frac{1}{20} =$ $\frac{6}{20} + \frac{18}{20} + \frac{1}{20} = \frac{25}{20} = 1\frac{5}{20} = \left(1\frac{1}{4}\right)$

5. $\frac{7}{16} + \frac{3}{32} + \frac{1}{4} =$

6. $1\frac{3}{4} + \frac{7}{8} + 1\frac{1}{16} =$

7. $\frac{5}{32} + \frac{7}{64} + \frac{7}{8} =$

8. $1\frac{3}{8} + \frac{3}{32} + \frac{7}{16} =$

9. $3\frac{1}{16} + \frac{9}{16} + \frac{1}{2} =$

10. $5 \frac{1}{5} + 2 \frac{3}{10} + 8 \frac{1}{2} =$

11. $4 \frac{5}{8} + 20 \frac{7}{32} =$

12. $\frac{3}{8} + \frac{7}{64} + \frac{9}{16} =$

13. $12 \frac{7}{8} + 25 \frac{3}{8} =$

14. $\frac{21}{32} + \frac{9}{64} + \frac{1}{4} =$

15. $\frac{3}{8} + 1 \frac{1}{2} + \frac{7}{16} + \frac{7}{8} =$

16. $2 \frac{1}{4} + \frac{5}{8} + \frac{5}{16} + \frac{17}{32} =$

Subtraction of Fractions

To subtract common fractions, the denominators must all be the same.

Example: $\frac{3}{4} - \frac{5}{16} = ?$

The LOWEST COMMON DENOMINATOR into which these denominators can be divided is 16.

Change fractions to a higher form (Fundamental Step No. 5)

$$\frac{3}{4} \times \frac{4}{4} = \frac{12}{16}$$

Subtract the numerators

$$\frac{12}{16} - \frac{5}{16} = \frac{7}{16}$$

Construction Mathematics Activity 3–2
SUBTRACTION OF FRACTIONS

Solve the following problems. Reduce answers to lowest form.

1. $\frac{3}{8} - \frac{1}{4} =$

2. $\frac{3}{4} - \frac{5}{16} =$

3. $1 \frac{7}{8} - \frac{13}{16} =$ $\frac{15}{8} - \frac{13}{16} = \frac{30}{16} - \frac{13}{16} = \frac{17}{16} = 1 \frac{1}{16}$

4. $3 \frac{1}{2} - \frac{9}{16} =$

(borrow $\frac{16}{16}$ from 3)

5. $10 \frac{3}{8} - 7 \frac{3}{32} =$

6. $5 - 2 \frac{3}{8} = \frac{5}{1} - \frac{19}{8} = \frac{40}{8} - \frac{19}{8} = \frac{21}{8} = 2 \frac{5}{8}$

7. $12 \frac{1}{16} - 8 \frac{1}{2} =$

8. $4 \frac{1}{4} - 3 \frac{1}{16} =$

9. $20 \frac{7}{8} - 11 \frac{3}{64} =$

10. $15 \frac{5}{8} - 5 \frac{1}{2} =$

Multiplication of Fractions

Common fractions may be multipled as follows:

1. Change all mixed numbers to improper fractions.

2. Multiply all numerators to get the numerator part of the answer.

3. Multiply all denominators to get the denominator part of the answer.

4. Reduce the fraction to lowest form..

Example: $\frac{1}{2} \times 3 \frac{1}{8} \times 4 = ?$

$$\frac{1}{2} \times \frac{25}{8} \times \frac{4}{1} = \frac{100}{16}$$

$$\frac{100}{16} = 6 \frac{4}{16} = 6 \frac{1}{4}$$

Construction Mathematics Activity 3–3
MULTIPLICATION OF FRACTIONS

Solve the following problems. Reduce answers to lowest form.

1. $\frac{3}{4}$ x $\frac{1}{2}$ = $\frac{3}{8}$

1. $2\frac{3}{4} \div 6 =$ $\frac{11}{4} \times \frac{1}{6} = \frac{11}{24}$

2. $2\frac{5}{8}$ x $\frac{1}{4}$ = $\frac{21}{8} \times \frac{1}{4} = \frac{21}{32}$

2. $12 \div \frac{3}{4} =$ $\frac{12}{1} \times \frac{4}{3} = 16$

3. $\frac{7}{8}$ x 5 = $\frac{35}{8} = 4\frac{3}{8}$

3. $16\frac{1}{8} \div 2 =$

4. $6\frac{3}{4}$ x $\frac{1}{3}$ = $\frac{27}{4} \times \frac{1}{3} = \frac{9}{4} = 2\frac{1}{4}$

4. $8\frac{2}{3} \div \frac{1}{3} =$ $\frac{26}{3} \times \frac{3}{1} = 26$

5. $12\frac{1}{2}$ x $\frac{1}{2}$ =

5. $16\frac{1}{4} \div 20 =$

6. $4\frac{3}{4}$ x $\frac{1}{2}$ x $\frac{1}{8}$ =

6. $\frac{7}{8} \div \frac{7}{16} =$

7. 16 x $\frac{3}{4}$ =

7. $15 \div 1\frac{1}{4} =$

8. $9\frac{5}{8}$ x $\frac{1}{2}$ =

8. $21\frac{3}{8} \div 3\frac{1}{8} =$

9. 10 x $\frac{4}{5}$ =

9. $5\frac{1}{4} \div \frac{3}{8} =$

10. $\frac{14}{3}$ x 6 =

10. $3\frac{5}{8} \div 2 =$

Division of Fractions

Common fractions may be divided as follows:

1. Change all mixed numbers to improper fractions.

2. Invert (turn upside down) the divisor and proceed as in multiplication.

Example: $5\frac{1}{4} \div 1\frac{1}{2} = ?$

$$\frac{21}{4} \div \frac{3}{2} =$$

$$\frac{21}{4} \times \frac{2}{3} = \frac{42}{12}$$

$$\frac{42}{12} = 3\frac{6}{12} = 3\frac{1}{2}$$

Construction Mathematics Activity 3—4
DIVISION OF FRACTIONS

Solve the following problems. Reduce answers to lowest form.

Decimal Fractions

The denominator in decimal fractions is 10 or a multiple of 10 (100, 1000, etc.). When writing decimal fractions, we omit the denominator and place a decimal point in front of the numerator:

$\frac{3}{10}$ is written .3 (three tenths)

$\frac{87}{100}$ is written .87 (eighty seven hundredths)

$\frac{375}{1000}$ is written .375 (three hundred seventy five thousandths)

$\frac{4375}{10000}$ is written .4375 (four thousand three hundred seventy five ten thousandths)

Whole numbers are written to the left of the decimal point and fractional parts are to the right:

$5\dfrac{253}{1000}$ is written 5.253 (five and two hundred fifty three thousandths)

Addition and Subtraction of Decimals

Decimals are added and subtracted in the same manner as whole numbers. In decimals, however, we write the figures so that the decimal points line up vertically.

Example:

```
Add    7.3125        Subtract 8.625
       1.25                   2.25
        .625                  6.375
       3.375
      12.5625
```

The decimal point in the answer is directly below the decimal points in the problem.

Construction Mathematics Activity 3−5
ADDITION AND SUBTRACTION OF DECIMALS

Solve the following problems:

Add:
```
       1. 4.5625
           .875
          2.75
        + 5.8137
```

2. 1.9375 + 3.25 + .375

3. 7.0625 + .125 + 8.0

4. 11.342 + 16.17 + .4207

5. .832 + .4375 + .27

Subtract:
```
       6. 27.9375
        − 16.937
```

7. 3.306 − 1.875

8. 4.0 − .0625

9. 10. − .75

10. 2.25 − 1.125

Multiplication of Decimals

Decimals are multiplied in the same manner as whole numbers, and the decimal points are disre-

garded until the multiplication is completed. To find the position of the decimal point in the answer: count the total number of decimal places to the right of the decimal point in the numbers being multiplied; then set off this number of decimal places in the answer, starting at the right.

Example:

```
   6.25  ⎱
 x 1.5   ⎰  3 decimal places
   3125
   625
   9.375  ⎱  3 decimal places
```

Construction Mathematics Activity 3−6
MULTIPLICATION OF DECIMALS

Solve the following problems:

```
1.    4.825
    x 1.75
```

```
2.   12.05
   x 4.124
```

```
3.   167
   x .25
```

```
4.   .838
   x 5.9
```

```
5.   65.96
    x .37
```

6. .375 x 6

7. 4.95 x 1.35

8. 3.75 x 100

9. 93.18 x .07

10. 5639.25 x 10

Division of Decimals

In dividing decimals, proceed as in the division of whole numbers except that the decimal point must be properly placed in the quotient (answer).

To place the decimal point in the quotient, count the number of places to the right of the decimal point in the divisor. Add this number of places to the right of the decimal point in the dividend and place the decimal point directly above in the quotient.

Example: $36.5032 \div 4.12 = ?$

$$
\begin{array}{r}
8.86 \\
4.12 \enclose{longdiv}{36.5032} \\
32\ 96 \\
\hline
3\ 543 \\
3\ 296 \\
\hline
2472 \\
2472 \\
\end{array}
$$

Divisor   4.12   Dividend

Construction Mathematics Activity 3—7
DIVISION OF DECIMALS
Solve the following problems:

1. $2.7\,\overline{)\,9.45}$

2. $.96\,\overline{)\,7.9392}$

3. $35\,\overline{)\,654.5}$

4. $2.4\,\overline{)\,172.8}$

5. $1.65\,\overline{)\,1386,0}$

6. $25924.64 \div 31.6$

7. $331.266 \div 80.6$

8. $821.7 \div 83$

9. $4401.25 \div 503$

10. $2585.52 \div 26.6$

Area Measurement

Frequently, it is necessary to know the amount of floor or wall space in a particular room or building. This measurement is known as area, and it is given in square units (square feet, square yards or square meters).

SQUARE AND RECTANGULAR AREAS. To figure the areas of a floor if it is rectangular or square in shape, multiply one side times an adjacent side (length x width). The product is the square foot area of the floor. See Fig. 3-2 (a).

Example:

$$
\begin{array}{r}
12 \\
\times\ 10 \\
\hline
120 \text{ square feet}
\end{array}
$$

The area of a wall section would be figured in the same way, except that the area of all openings (doors, windows, etc.) would be subtracted from the total. See Fig. 3-2 (b).

Example:

$$
\begin{array}{r}
20 \\
\times\ 8 \\
\hline
160 \text{ sq. ft.}
\end{array}
$$

$$
\begin{array}{r}
5 \\
\times\ 4 \\
\hline
20 \text{ sq. ft.}
\end{array}
$$

$$
\begin{array}{r}
160 \\
-20 \\
\hline
140 \text{ square feet of wall surface}
\end{array}
$$

TRIANGULAR AREA. An area such as a triangular gable may be figured by multiplying the height times one-half of the base. Fig. 3-3 explains why the formula is 1/2 B x H.

Example of figuring a gable end area is as follows:

$$
\begin{array}{r}
12 \\
\times\ 5 \\
\hline
60 \text{ square feet}
\end{array}
$$

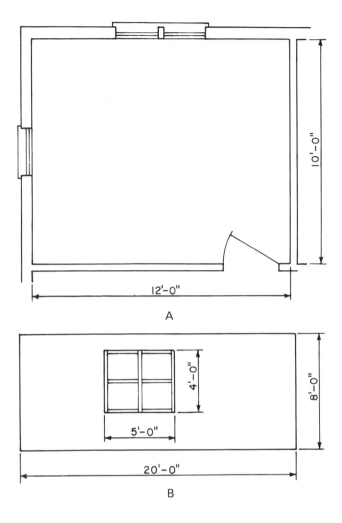

A

B

Fig. 3-2. A—Formula for finding area of a floor is: A = L x W. B—Area of a wall is : A = L x W — area of openings.

CIRCULAR AREA. The area (A) of a circle may be found by using the formula pi (3.1416) times radius squared: $\pi r^2 = A$

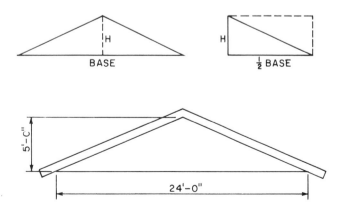

Fig. 3-3. Formula for finding area of a triangular gable is: A = 1/2B x H.

Example shown in Fig. 3-4.

Patio diameter = 30 feet, radius = 15 feet

3.1416 x 15 x 15 = A

3.1416 x 225 = A

$$\begin{array}{r} 3.1416 \\ \times\ \ 225 \\ \hline 157080 \\ 62832\ \ \\ 62832\ \ \ \ \\ \hline \end{array}$$
706.8600 square feet in patio

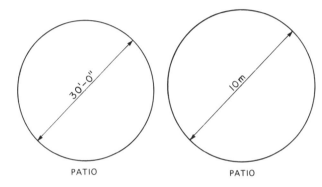

PATIO   PATIO

Fig. 3-4. Formula for finding area of a circle is: $A = \pi r^2$.

Volume Measurement

In some jobs, such as concrete work, volume measure must be figured. Volume is cubic measure. It is found by multiplying area (square measure) by depth. Referring to the patio used as an example, we will figure the volume of ready-mix concrete required for a 4 inch slab.

To get volume, you must multiply square feet times feet in depth to get cubic feet. Therefore, change the 4 inches to .333 feet.

$$\begin{array}{r} 706.86\ \text{sq. ft. (area of patio)} \\ \times\ .333 \\ \hline 212058 \\ 212058\ \ \\ 212058\ \ \ \ \\ \hline \end{array}$$
235.38438 cubic feet in patio slab

Since concrete is sold by the cubic yard and not by cubic feet, it is necessary to change the cubic feet to cubic yards. There are 27 cubic feet in a cubic yard, so we divide cubic feet by 27.

Example:

$$27 \overline{)235.38} \qquad 8.717 \text{ cubic yards of concrete in patio}$$

```
         8.717 cubic yards of concrete in patio
  27 |235.38
      216
      ‾‾‾‾‾
      193
      189
      ‾‾‾‾
       48
       27
      ‾‾‾‾
      190
      189
      ‾‾‾‾
        1
```

Area and volume measurements in metric are figured in the same manner using appropriate units of centimeters (cm) or meters (m).

Example:

Patio diameter = 10 meters; thickness = 10 centimeters.

```
   3.1416
 x    25 (radius squared)
 ‾‾‾‾‾‾‾
 157080
  62832
 ‾‾‾‾‾‾‾
 78.5400  = area of patio in square meters (m²)
```

```
  78.54
 x 0.1 m (10 cm = 0.1 m)
 ‾‾‾‾‾
  7.854 = volume of patio in cubic meters (m³)
```

Note that fewer calculations are required when units are in metric.

Construction Mathematics Activity 3—8
PROBLEMS IN CONSTRUCTION MATHEMATICS

Solve the following problems. Show your work.

1. A triangular frame has sides that measure 15.7, 20.4 and 26.2 centimeters. What is the total length of the three sides?

2. A carpenter had a board 34 3/4 inches long. To fit the space for a shelf, he cut 7/16 inch off one end. How long was the board after the piece was removed.

3. Fifteen strips, 1 1/4 inches wide, are to be ripped from a sheet of plywood. If 1/8 inch is lost with each cut, how much of the plywood sheet is used in making the 15 strips? (Assume 15 cuts are necessary.)

4. An interior wall of a house is made up of 2 x 4 studs with 5/8 inch wallboard on each side. See the drawing below. If the actual width of a 2 x 4 stud is 3 3/8 inches, what is the total thickness of the wall?

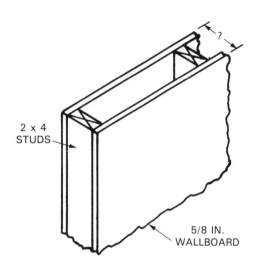

2 x 4 STUDS

5/8 IN. WALLBOARD

5. A carpenter worked 10 weeks on a particular job, 5 1/2 days per week and 7 3/4 hours per day. How many hours did he work on the job?

6. What is the area of a rectangular floor that is 7.3 meters long and 4.2 meters wide?

7. There are 15 risers in a set of stairs running from the basement to the first floor. Each riser is 7 1/4 inches high. What is the distance between floors?

8. The distance between two floors is 108 1/2 inches. If 14 risers are to be used in a set of stairs, what is the height of each riser?

9. How many 1 x 2 shelf cleats 3/4 of a foot long can be cut from a 1 x 2 board 16 feet long? How much is left? (Disregard waste in saw cut.)

10. A contractor removed 35.7 cubic meters (m³) of earth from a building site. If his trucks can haul 1.7 cubic meters per load, how many truck loads of earth were moved?

11. Figure the amount of concrete required to pour a floor slab of the following dimensions: 18 feet by 24 feet by 4 inches.

12. How much concrete is needed to pour a slab 6 meters by 8 meters by 10 centimeters?

13. How much paint is required to paint one side of a block wall of the following dimensions: 6 feet by 172 feet? The paint being used will cover 200 square feet per gallon.

14. Figure the amount of sealer required to seal a floor 12 meters by 25 meters. The sealer being used will cover 7 square meters per liter.

15. How much concrete is required to pour the slab shown below?

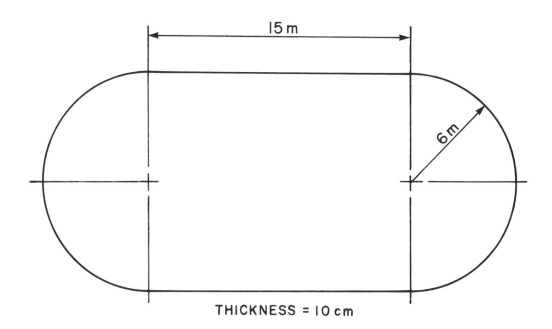

THICKNESS = 10 cm

PART 2

BLUEPRINT READING FUNDAMENTALS

Unit 4
Alphabet of Lines and Symbols

The reading and understanding of construction blueprints begins with recognition of lines and symbols that appear on drawings. Drafters use a definite system of lines and symbols. How they are drawn and the meaning of each line and symbol are presented in this unit.

The Alphabet of Lines

Nine lines are commonly used on construction drawings. These are referred to as the "Alphabet of Lines." When properly used, each of these lines helps convey meaning to the drawing. The lines vary in "weight" or thickness and may be a solid line or a combination of broken lines. Fig. 4-1 illustrates these lines and shows examples of how they are used.

PROPERTY LINE: The property line is an extra heavy line with long dashes alternating with two short dashes, Fig. 4-1. The length and bearing (direction) of each line usually is identified on the site plan.

MAIN OBJECT LINE: Object lines represent the main outline of the features of the object, building or walk. The object line is a heavy, continuous line, showing all edges and surfaces, Fig. 4-1.

HIDDEN LINE: Hidden lines are medium weight, short dashes, showing edges and surfaces which are not visible in a particular view, Fig. 4-1. The worker must look for another view in the set of drawings to locate at what location these edges occur. Often these hidden parts will be revealed in an elevation or in a sectional view. Hidden lines are used only when their presence helps to clarify a drawing.

CENTER LINE: The center line is used to indicate centers of objects such as columns, equipment and fixtures. These objects usually are located by dimensioning to the center. The center line is also used to indicate the finished floor line. This line is light in weight and composed of alternating long and short dashes, Fig. 4-1. Sometimes, the symbol ℄ is used to identify the center line.

DIMENSION AND EXTENSION LINE: Dimension and extension lines are thin lines to indicate the extent and direction of dimensions. Dimension lines usually are terminated against extension lines with arrowheads, slashes or dots, Fig. 4-1. Leaders are the same weight as dimension and extension lines. Leaders are terminated with an arrowhead or dot against the feature on a drawing to which a note or other reference is directed, Fig. 6-11.

BREAK LINES: Break lines are used to limit a partial view to indicate that the object continues but is not shown on this drawing; or to indicate that the full length of the object has not been drawn to save space. When the break on the drawing is lengthy, LONG BREAK lines with freehand "zig-zags" are used. See Fig. 4-1. SHORT BREAK lines are used when the break is short, such

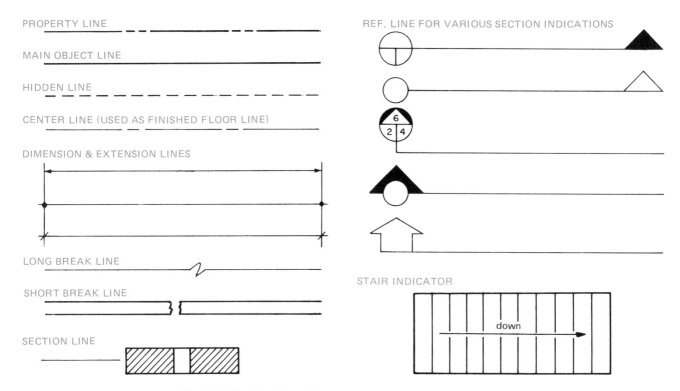

Fig. 4-1. The Alphabet of Lines used on construction drawings.

as across a joist or beam. This is a thick line drawn freehand. Some architectural draftsmen use the long break line for all breaks.

SECTION LINE: Section lines, sometimes referred to as cross-hatch lines, are thin lines usually drawn at an angle of 45 degrees. They are used to show the interior detail of a wall or other structure. See Fig. 4-1.

REFERENCE LINE FOR SECTION: Reference lines are solid lines indicating that an imaginary cut has been made at this point, and that a detail section is shown elsewhere on the drawings, Fig. 4-1. The arrow indicates the direction in which the section is viewed. The letters and numerals, usually in a circle attached to the reference line, indicate the particular section and where it will be found. For example, the reference may read: "Section No. 6 on Sheet No. 2 and is shown on Sheet No. 4 of the set of blueprints."

STAIR INDICATOR: The ascent or descent of a run of stairs is indicated by a short light line with

an arrow head at one end. The direction the stairs run on the particular plan view is indicated by the word "up" or "down" which is placed above the stair indicator line.

Symbols Used on Construction Drawings

In addition to the various lines that give meaning to a drawing, a number of symbols (sometimes referred to as conventions) are commonly used on construction drawings.

Two types of symbols are used: One type consists of picture-like representations that are easily recognized, such as electrical symbols. See page 160. The other type represents materials that are not recognizable unless you know the symbol. See page 148. Typical symbols used on construction drawings are shown on pages 76, 84, 101, 144, 171, 238 and 240. Regardless of the construction field of work in which you are employed, you should be familiar with all the symbols for they may affect your area of construction. Study the symbols in this text.

Alphabet of Lines and Symbols

Line Activity 4−1
ALPHABET OF LINES

Draw freehand, in the spaces provided, the various lines used in construction blueprints. Pay close attention to the FORM and WEIGHT of each line.

Example:

1. Main Object Line

2. Property Line

3. Hidden Line

4. Center Line

5. Dimension and Extension Line

6. Broken Line

7. Section Line

8. Reference Line for Section

Symbols Activity 4−2
CONSTRUCTION DRAWING SYMBOLS

Draw freehand in the spaces provided, the symbol for each item listed. Refer to pages 76, 84, 101, 144, 148, 160, 171, 238 and 240 for correct symbols.

1. Duplex Convenience Outlet

5. Three-Way Switch

2. Push Button

6. Shower Stall

3. Floor Outlet

7. Floor Drain

4. Telephone Outlet

8. Water Heater

9. Hot Water Line

10. Water Closet

11. Concrete, Cast

12. Concrete, Block

13. Framing Lumber

14. Insulation Batts or Fill

15. Brick

16. Steel Reinforcing Rod

17. Steel Angle Plate

18. Standard Beam (I)

19. Supply Duct

20. Glass, Elevation

Unit 5
Freehand Technical Sketching

Sketching is a technique frequently used by craftsworkers, suppliers, architects and engineers to communicate ideas and construction details to others. This unit presents the basics of sketching as a means of helping you learn to read and interpret blueprints.

Several types of papers are suitable for sketching, including plain note paper and cross section paper. In applying the sketching techniques discussed in this text, use plain paper without ruling. Then, when you have developed the basic techniques of sketching on plain paper, you will have no difficulty using cross section paper.

Sketching Technique

Sharpen your pencil to a conical point for sketching as shown in Fig. 5-1.

The manner in which the pencil is held in freehand sketching is important. Grip it firmly enough to control the strokes but not so tight as to stiffen your movements. Your arm and hand should have a free and easy movement. The point of the

pencil should extend approximately 1 1/2 inches beyond your fingertips, Fig. 5-2.

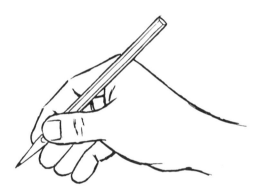

Fig. 5-2. Holding pencil for sketching.

Rotate your pencil slightly between strokes to maintain the point. Initial lines should be firm and light but not fuzzy. Avoid making grooves in your paper caused by using too much pressure. In sketching straight lines, your eye should be on the point at which the line is to terminate. Make a series of short strokes (lines), rather than one continuous stroke. See Fig. 5-3.

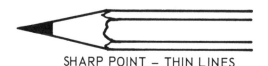

SHARP POINT – THIN LINES

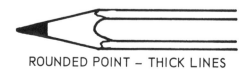

ROUNDED POINT – THICK LINES

Fig. 5-1. Pencil points for sketching.

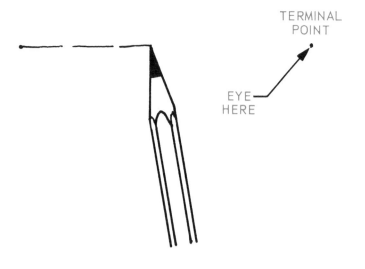

Fig. 5-3. Sketching straight lines.

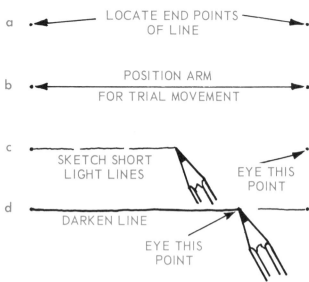

Fig. 5-5. Steps in sketching horizontal lines.

Sketching Horizontal Lines

Horizontal lines are sketched with a movement of the forearm approximately perpendicular to the line being sketched, Fig. 5-4.

Four steps are essential in sketching horizontal lines:

1. Locate and mark the end points of line to be sketched, Fig. 5-5(a).
2. Position your arm by making trial movements from left to right (left-handers, from right to left) without marking the paper, Fig. 5-5(b).

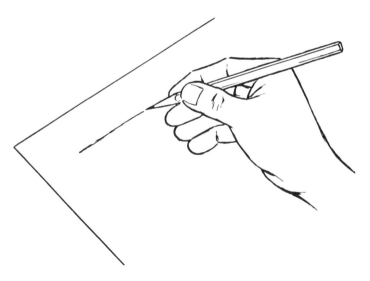

Fig. 5-4. Position for sketching horizontal lines.

3. Sketch short, light lines between the points, Fig. 5-5(c). KEEP YOUR EYE ON THE POINT WHERE LINE IS TO END.

4. Darken the line to form one continuous line of uniform weight. The eye should lead the pencil along the lightly sketched line. See Fig. 5-5(d).

Sketching Assignment

Turn to page 36 and sketch the series of horizontal lines called for in Sketching Assignment 5 – 1. Follow the suggested steps of procedure closely and work for improvement with each line.

Sketching Vertical Lines

Vertical lines are sketched from top to bottom, using the same short strokes in series as for horizontal lines. When making the strokes, position your arm comfortably at approximately 15 deg. with the vertical line, Fig. 5-6. A finger and wrist movement together with a pulling arm movement are best for sketching vertical lines.

These four steps should be used in sketching vertical lines:

1. Locate and mark the end points of the line to be sketched, Fig. 5-7(a).
2. Position your arm by making trial movements from top to bottom without marking the paper, Fig. 5-7(b).
3. Sketch short, light lines between the points, Fig. 5-7(c). KEEP YOUR EYE ON THE POINT WHERE LINE IS TO END.

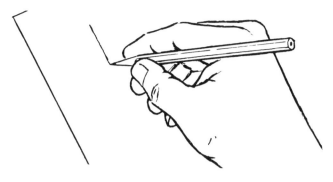

Fig. 5-6. Position for sketching vertical lines. Paper may be rotated slightly counterclockwise for greater ease.

4. Darken the line to form one continuous line of uniform weight. The eye should lead the pencil along the lightly sketched line. See Fig. 5-7(d).

You may find it easier to sketch horizontal and vertical lines if the paper is rotated slightly counter-clockwise, Fig. 5-6.

Sketching Assignment

Turn to page 36 and sketch the series of vertical lines called for in Sketching Assignment 5 – 2. Follow the suggested steps closely and work for improvement with each line.

Sketching Inclined Lines and Angles

All straight lines which are neither horizontal nor vertical are called INCLINED LINES. You can sketch the inclined lines between two points or at a

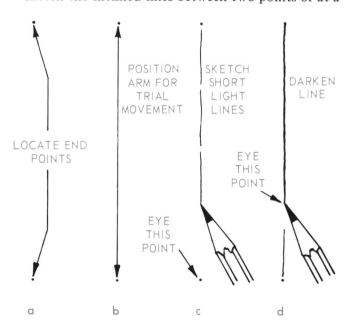

Fig. 5-7. Steps in sketching vertical lines.

designated angle. The same strokes and techniques used for sketching horizontal and vertical lines are used for inclined lines or angles, depending on their position. If you prefer, rotate the paper to sketch these lines horizontally or vertically.

Angles may be estimated quite closely by first sketching a right angle (90 deg.), then subdividing its arc to get the desired angle. Fig. 5-8 illustrates how this is done to get an angle of 30 deg.

Sketching Assignment

Turn to page 37 and sketch the angles called for in Sketching Assignment 5 – 3.

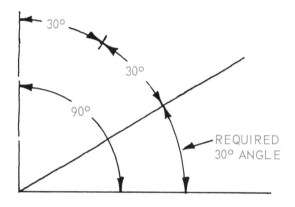

Fig. 5-8. Estimating angle sizes in sketching.

Sketching Arcs and Circles

There are several methods of sketching arcs and circles but the triangle-square method usually is most satisfactory.

To SKETCH AN ARC connecting two straight lines:
1. Project the two lines until they intersect, Fig. 5-9(a).
2. Lay out desired arc radius from the point of the intersecting lines, Fig. 5-9(b).
3. Form a triangle by connecting these two points, then locate the center point of the triangle. See Fig. 5-9(c).
4. Sketch short, light strokes from the point where the arc is to start on the vertical line through the center point to the point on the horizontal line where the arc ends, Fig. 5-9(d).
5. Darken the line to form one continuous arc,

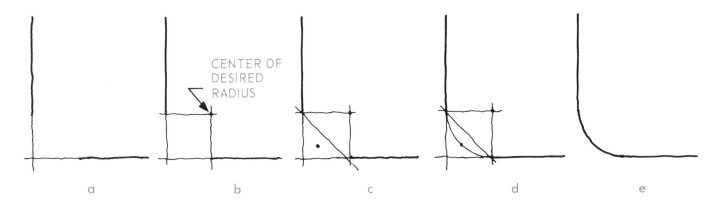

CENTER OF
DESIRED
RADIUS

a   b   c   d   e

Fig. 5-9. Steps in sketching an arc.

which should join smoothly with each straight line. Erase construction lines, Fig. 5-9(e).

To SKETCH A CIRCLE of a certain diameter:
1. Locate the center of the circle and sketch the center lines. Then, lay off half the diameter on each side of the two diameters. See Fig. 5-10(a).
2. Sketch a square lightly at the diameter ends, Fig. 5-10(b).
3. Across each corner, sketch a diagonal line to form a triangle. Then, locate the center point of each triangle, Fig. 5-10(c).
4. Sketch short, light strokes through each quarter of the circle, making sure the arc passes through the triangle center point and joins smoothly with the square at the diameter ends. See Fig. 5-10(d).
5. Darken the line to form a smooth, well-formed circle. Erase construction lines, Fig. 5-10(e).

Sketching Assignment

Turn to page 38 and sketch the arc and circles called for in Sketching Assignment 5−4. Follow the suggested steps just discussed for sketching arcs and circles.

To SKETCH AN ELLIPSE of a certain major and minor axis:
1. Locate center of ellipse and sketch center lines, Fig. 5-11(a).
2. Lay off major axis of ellipse on horizontal center line and minor axis on vertical center line. See Fig. 5-11(b).
3. Sketch rectangle through points on axis, Fig. 5-11(c).
4. Sketch tangent arcs at points where center lines cross rectangle, Fig. 5-11(d).
5. Complete the ellipse and darken, then erase the construction lines. See Fig. 5-11(e).

Proportion in Sketching

Proportion in sketching is the relationship of the size of one part to another and to the object as a whole. The width, height and depth of your sketch must be kept in the same proportion so the sketch conveys an accurate description of the object being sketched.

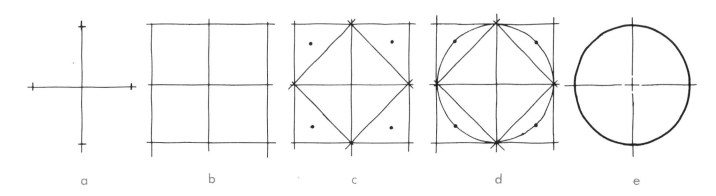

a   b   c   d   e

Fig. 5-10. Steps in sketching a circle.

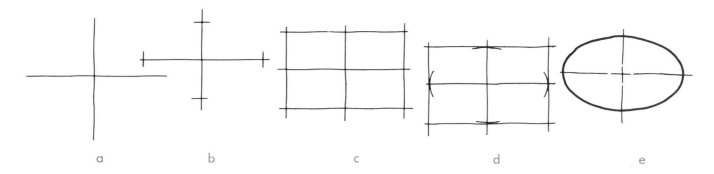

Fig. 5-11. Steps in sketching an ellipse.

One technique useful in estimating proportions is the UNIT method. This method involves establishing a relationship between measurements on the object by breaking each of the measurements into units. Compare the width to the height, and select a unit that will fit each measurement, Fig. 5-12. Lengths laid off on your sketch should be in the same proportion, although the units on the sketch may vary in size from those of the actual object.

Proportion is a matter of estimating lengths on a part or assembly, then setting these down on your sketch in the same ratio of units. Practice this

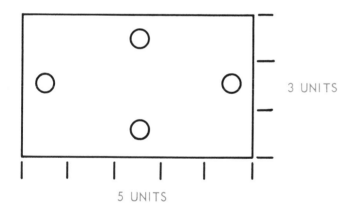

3 UNITS

5 UNITS

Fig. 5-12. Unit method of proportioning.

method of establishing proportion in sketches. It will help you develop skill in accurately representing objects you sketch.

Sketching Assignment

Turn to page 39 and sketch the swimming pool, Sketching Assignment 5−5. Follow the suggestions given to obtain good proportion.

Steps in Sketching an Object

The following steps will help you in laying out and completing your freehand sketches:
1. Sketch a rectangle, square, etc., of the correct proportion, Fig. 5-13(a).
2. Sketch major subdivisions and details of the object, Fig. 5-13(b).
3. Remove unnecessary lines with eraser, Fig. 5-13(c).
4. Darken lines to right weight, Fig. 5-13(d).

Aids to Freehand Sketching

Aids to freehand sketching include cross section paper for making sketches and enclosing squares and rectangles for drawing circles and ellipses. There are other aids, such as using a piece of folded paper, cardboard or 6-inch rule for a straightedge, and a scrap of paper for measuring proportion or in laying off the radius of a circle.

The aim in freehand sketching should be to develop your skill to a point where aids will no longer be necessary. Sketching is a means to quickly and effectively communicate the shape and size descriptions of an object to other persons; or, to record information in the field for use at a later time.

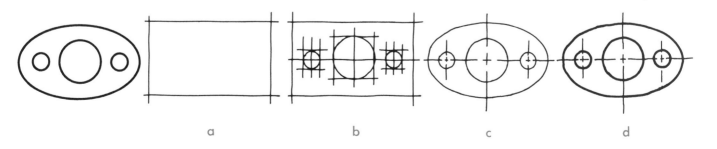

Fig. 5-13. Steps in sketching an object.

Sketching Assignment 5—1
HORIZONTAL LINES

Sketch horizontal lines between points A—A' through J—J'.

A·   ·A' B·   ·B'

C·   ·C' D·   ·D'

E·   ·E'

F·   ·F'

G·   ·G'

H·   ·H'

I·   ·I'

J·   ·J'

Sketching Assignment 5—2
VERTICAL LINES

Sketch vertical lines between points K—K' through T—T'.

K   M   O   P   Q   R   S   T

K'   M'
L   N

L'   N'   O'   P'   Q'   R'   S'   T'

Sketch the required angles starting at the indicated point.

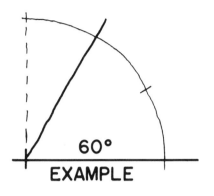

60°
EXAMPLE

45°
A

75°
B

20°
C

100°
D

120°
E

$22\frac{1}{2}°$
F

85°
G

50°
H

30°
I

15°
J

90°
K

Sketching Assignment 5—4
ARCS AND CIRCLES

Sketch arcs joining the sets of lines A through F. Show construction lines for A through C. Erase construction lines for D through F. Sketch circles G through L. Show construction lines for circles G through I. Erase construction lines for J through L.

A

B

C

D

E

F

G

H

I

J

K

L

Sketch the layout for the swimming pool in the space below the drawing. Estimate the proportions. Do not measure or dimension the sketch. Date the sketch and sign your name.

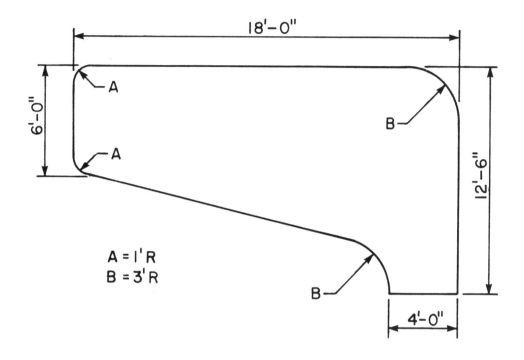

Sketching Assignment 5–6
FLOOR PLAN

Sketch the floor plan in the space below the drawing. Estimate the proportions. Do not measure or dimension the sketch. Date the sketch and sign your name.

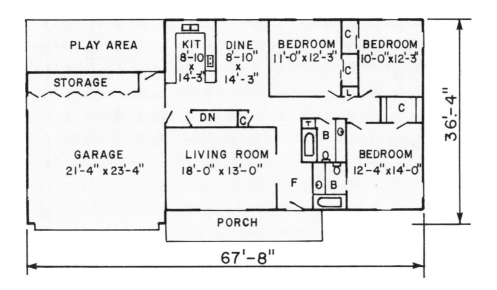

Unit 6
Pictorial Drawings

Pictorial drawings show a room, a cabinet or a complete building much as it would appear in a photograph, or as if you were viewing the actual object. See Fig. 6-1. Several sides of the object are visible in one composite view.

Pictorial drawings are quite easy to understand. They give an overall view of a room or structure, showing relationship in construction or assembly. Therefore, you will find this unit on the technique of pictorial sketching useful in helping you to visualize two and three view orthographic (ortho-graf-ick) drawings, as well as in communicating your ideas on technical problems to others.

There are three common types of pictorial drawings in general use:
1. Isometric (i-so-met' rik).
2. Oblique (ob-lek').
3. Perspective (per-spek-tive).

A fourth type is the exploded pictorial drawing which is a special use of one of the three more common types.

Isometric Drawing

Isometric drawings and sketches are used more often than other types of pictorials. They are easily constructed and measurements can be made directly on the view.

An isometric drawing is constructed with its two faces projected at angles of 30 deg. with the horizontal. See Fig. 6-2. The dark lines in the center are called the isometric axes. They are equally spread at 120 deg. between each set and use the same scale of measure along each axis. The word "isometric" means equal measure. In this type drawing, it means equal measure on all axes.

Lines that are horizontal in an orthographic drawing are drawn at an angle of 30 degrees in an isometric drawing. Vertical lines remain vertical, Fig. 6-2. Slant lines (non-isometric lines) are drawn by locating their end points on the isometric axes and connecting the two points.

Constructing an Isometric Sketch

Steps in the construction of an isometric sketch are shown in Fig. 6-3.

PROCEDURE FOR CONSTRUCTING AN ISOMETRIC SKETCH

1. Given three views of a sawhorse, Fig. 6-3(a).
2. Select the position of the object to best describe its shape.
3. Start the sketch by laying out the axes for the lower corner, Fig. 6-3(b).
4. Make overall measurements in their true length on the isometric axes or on lines parallel to the axes. See Fig. 6-3(c).
5. Construct a "box" to enclose the object, Fig. 6-3(d).
6. Sketch the isometric lines of the object, Fig. 6-3(e).

Fig. 6-1. A pictorial sketch of a kitchen. (St. Charles Manufacturing Co.)

Perspective View—Range Elevation

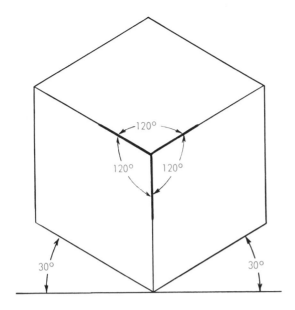

Fig. 6-2. Isometric axes.

7. Sketch the non-isometric lines (slant lines) by first locating the end points of these lines, then sketching the line between, Fig. 6-3(f).
8. Darken all visible lines and erase the construction lines to complete the isometric sketch. See Fig. 6-3(g).

Circles and Arcs in Isometric

Circles and arcs in isometric are sketched in the same manner you learned to sketch circles and arcs in Unit 5, except you start with an isometric square.

PROCEDURE FOR SKETCHING ISOMETRIC CIRCLES

1. Sketch an isometric square to enclose location of circle, Fig. 6-4(a).
2. Locate the midpoints of the sides of the isometric square and connect these midpoints, Fig.6-4(b).
3. Locate the midpoints of the triangles formed, then sketch isometric arcs through each to form an isometric circle, Fig. 6-4(c).
4. Erase construction lines and darken circle, Fig. 6-4(d).

Circles may be sketched in all three planes of the isometric drawing in the same manner, Fig. 6-5.

Isometric arcs may be sketched as shown in Fig. 6-6.

PROCEDURE FOR SKETCHING AN ISOMETRIC ARC

1. Lay off the radius of the arc from the corner, Fig. 6-6(a).
2. Draw a slant line connecting the two points,

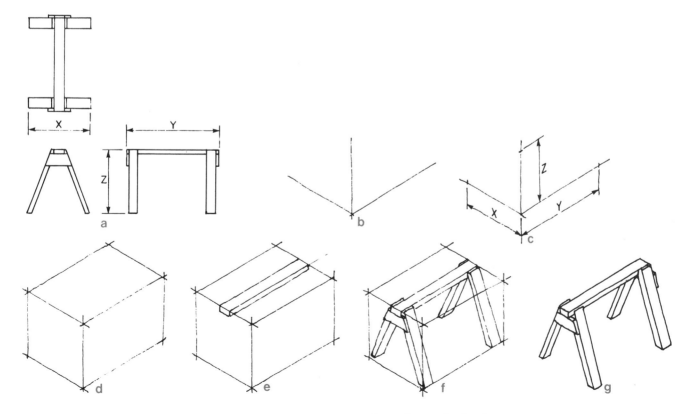

Fig. 6-3. Steps in constructing an isometric sketch.

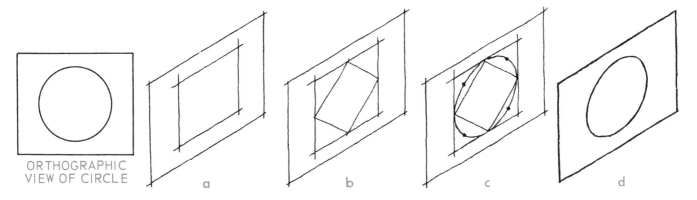

Fig. 6-4. Steps in sketching an isometric circle.

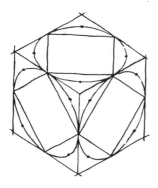

Fig. 6-5. Isometric circles in three planes.

lines are extended in line with these axes, Fig. 6-7. The dimension figures are usually aligned as shown in Fig. 6-7.

forming a triangle, Fig. 6-6(b).

3. Locate midpoint of triangle and sketch an arc through this point to join smoothly with sides, Fig. 6-6(c).
4. Erase construction lines and darken arc. See Fig. 6-6(d).

Isometric Dimensioning

The dimension lines on an isometric drawing or sketch are parallel to the isometric axes. Extension

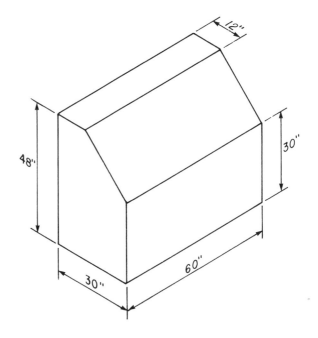

Fig. 6-7. Isometric dimensioning.

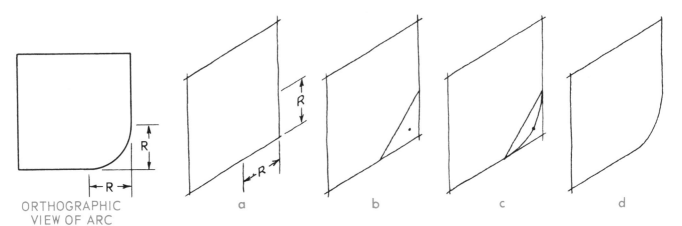

Fig. 6-6. Steps in sketching an isometric arc.

Make an isometric sketch of the open shelving wall unit. Omit dimensions.

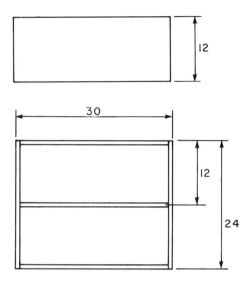

Sketching Assignment 6−2
CORNER WALL UNIT

Make an isometric sketch of the corner wall unit. Dimension your sketch.

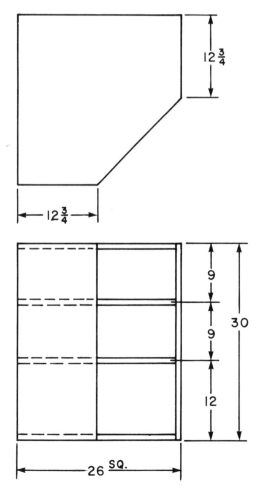

Oblique Drawing

Sometimes it is desirable to show the front view of a cabinet or other structure in its true shape and size (unless reduced or enlarged) just as in orthographic projection. This is an advantage when the front view contains circles or arcs, which can be represented as true circles and arcs. However, the top and side views are projected back from the front view at an angle, usually 45 degrees. This tends to give a distorted appearance to the drawing, shown in Fig. 6-8.

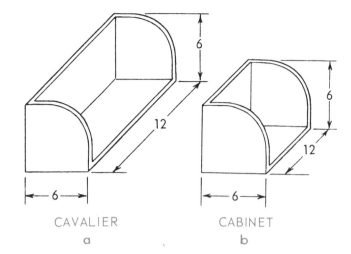

Fig. 6-9. Types of oblique drawings.

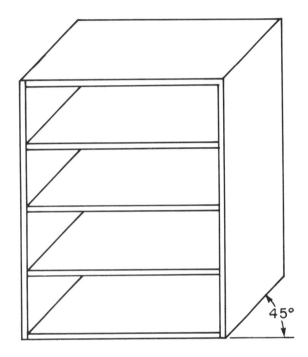

Fig. 6-8. An oblique drawing.

There are two principal types of oblique drawings: cavalier and cabinet.

Cavalier Oblique

Cavalier oblique drawings are drawn with their receding sides to the same scale as the front view, Fig. 6-9(a). This creates a severe distorted appearance, but it does have the advantage of using one scale throughout.

Cabinet Oblique

Cabinet oblique drawings differ from Cavalier drawings only by the fact that measurements made

on the receding axes are reduced by half, Fig. 6-9(b). This gives a much more pleasing appearance and is frequently used in the drawing of cabinet work.

Circles and Arcs in Oblique

Circles and arcs in oblique are sketched as true circles or arcs in the front plane as mentioned earlier, Fig. 6-10. When these occur in the receding planes, they are sketched in the same manner as isometric circles and arcs.

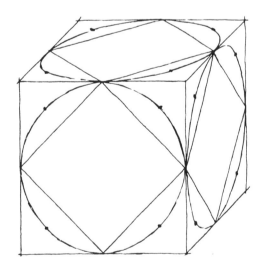

Fig. 6-10. Sketching oblique circles and arcs.

PROCEDURE FOR SKETCHING OBLIQUE CIRCLES AND ARCS

1. Sketch an oblique square to contain circle, Fig. 6-10.

2. Locate midpoints of sides of the oblique square and connect these midpoints.
3. Locate midpoints of triangles, then sketch oblique circles or arcs through these midponts to join smoothly with sides of oblique square.
4. Erase construction lines and darken circles and arcs.

The procedure for sketching circles and arcs is the same in cabinet oblique drawings as it is for cavalier oblique except the oblique square is reduced on the receding axes.

Oblique Dimensioning

Oblique dimensioning must be done in the same plane as the surface or feature appears, just as in isometric dimensioning. Fig. 6-9 illustrates how dimensions are placed on an oblique drawing.

Sketching Assignment 6 – 3
WALL UNIT

Make a cavalier oblique sketch of the wall unit (no shelf). Do not dimension.

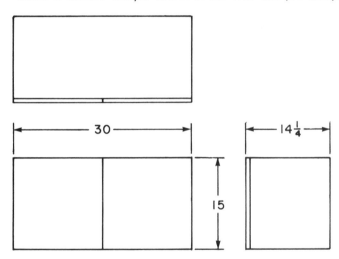

Sketching Assignment 6 – 4
WALL CUPBOARD

Make a cabinet oblique sketch of the wall cupboard (less doors).

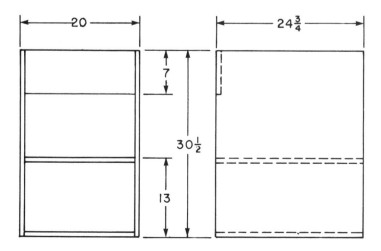

Perspective Drawing

The perspective drawing is the most realistic of all pictorial drawings. Instead of the receding lines remaining parallel (as in isometric and oblique drawings), receding lines in the perspective drawing or sketch tend to converge (meet at one point). See Fig. 6-11. This eliminates the distorted appearance that occurs at the back part of most other pictorial drawings.

To assist you in developing the technique of this type of sketching, both parallel and angular perspective sketches are discussed and illustrated.

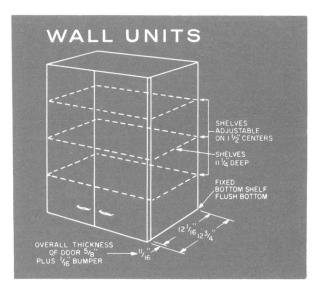

Fig. 6-11. Perspective drawing. (St. Charles Manufacturing Co.)

Parallel or Single-Point Perspective

When one face of an object appears in its true shape and size (unless reduced or enlarged), and it is parallel to the picture plane (as in orthographic projection), the perspective is known as parallel or single point perspective. That is, lines parallel to the front picture plane remain parallel, while the receding lines of the other two faces converge in the direction of a single vanishing point. See Fig. 6-12.

PROCEDURE FOR SKETCHING A PARALLEL OR SINGLE-POINT PERSEPECTIVE

1. Sketch the front view of the object in its true size and shape, as shown in an orthographic sketch in Fig. 6-13(a).

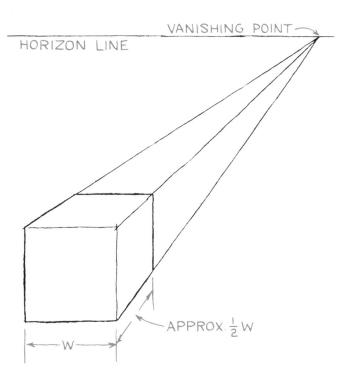

Fig. 6-12. Parallel or single-point perspective.

2. Sketch horizontal line (called horizon) at the assumed eye level of the viewer, Fig. 6-13(b). This line may be above, behind or below the object, depending on how you want to view the object.
3. Select vanishing point (VP) on the horizon as far to the right or left as desired. See Fig. 6-13(c).
4. Sketch lines from front view to VP, Fig. 6-13(d).
5. Enclose object in a "box" by sketching rear vertical and horizontal lines, Fig. 6-13(d). To estimate depth of side and top view, reduce these distances by about one-half and adjust until it pleases the eye.
6. To sketch slant lines, locate their end points on the perspective axes and sketch lines between. See Fig. 6-13(e).
7. Block in features such as drawers and doors, Fig. 6-13(f).
8. Darken visible lines and erase construction lines, Fig. 6-13(g).

Angular or Two-Point Perspective

The angular or two-point perspective gets its name from the fact that the two side faces of the object meet the front picture plane at an angle and recede toward two vanishing points on the horizon. See Fig. 6-14.

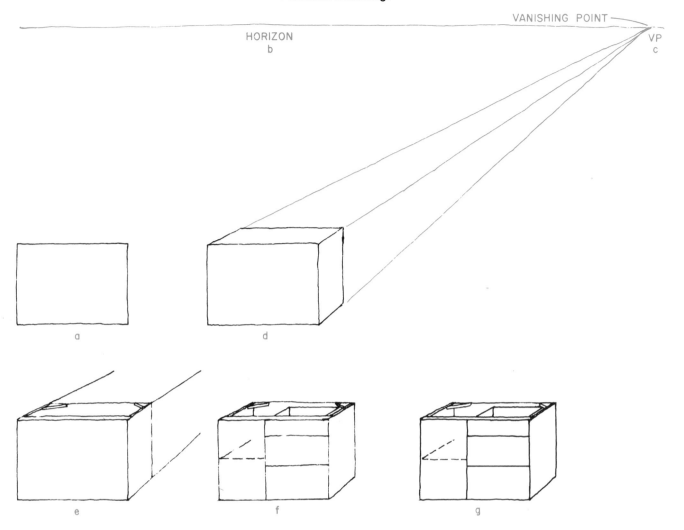

Fig. 6-13. Steps in sketching a parallel or single-point perspective.

PROCEDURE FOR SKETCHING AN ANGULAR OR TWO-POINT PERSPECTIVE

1. Sketch horizontal line (horizon) at the assumed eye level of the viewer, Fig. 6-14(a). This line may be above, behind or below the object, depending on the level from which you want to view the object. See Fig. 6-15 for perspective view with horizon below the object.
2. Select position in which object is to be viewed and sketch vertical line for front corner of a "box" to enclose object. See Fig. 6-14(b).
3. Establish right and left vanishing points on horizon, Fig. 6-14(c). If the object is to be positioned so that the two sides can be viewed equally, then the vanishing points will be equidistant on each side of the object. If one side is to be favored (front view in Fig. 6-14), the vanishing point for that side will be extended out while the vanishing point on the other side will be shortened. However, both VPs must remain on the line of the horizon.
4. Sketch receding lines from front vertical line to the two vanishing points, Fig. 6-14(d).
5. Enclose object in a "box" by sketching rear vertical lines, Fig. 6-14(e). To estimate depth of side and top views, reduce these distances by about one-half and adjust until it pleases the eye. The reduction is not as great for the side being favored. See the front view in Fig. 6-14.
6. Sketch slant lines by locating two end points on the perspective axes and sketch lines between, Fig. 6-14(f).
7. Block in features such as drawers and doors, Fig. 6-14(g).
8. Darken visible lines and erase construction lines.

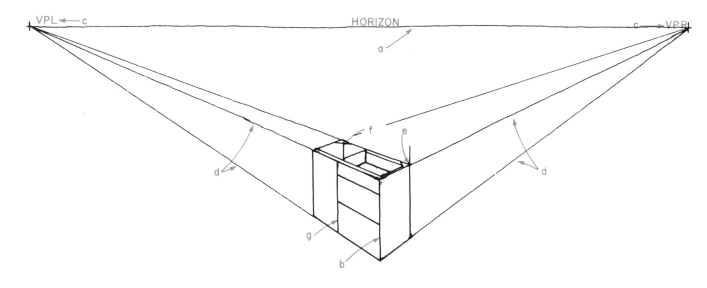

Fig. 6-14. *Steps in sketching an angular or two-point perspective.*

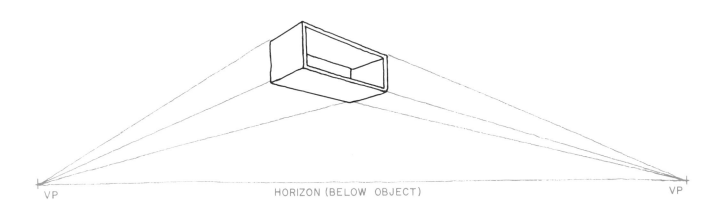

Fig. 6-15. *Perspective view of wall cabinet with horizon below object.*

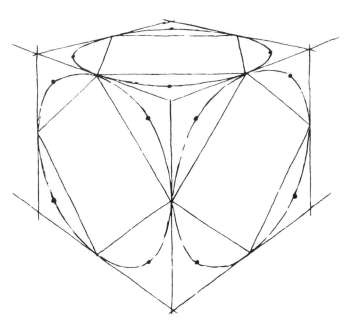

Fig. 6-16. *Sketching perspective circles and arcs.*

Circles and Arcs in Perspective

Circles and arcs in perspective are sketched in the same way as isometric circles and arcs by sketching first the perspective square or block, then joining the midpoints of the sides to form triangles. The perspective circle or arc is sketched through the midpoints of the sides and the center of the triangles. See Fig. 6-16.

Exploded Pictorial Drawings

Exploded view drawings are used to show relative position of parts or construction detail, Fig. 6-17. They are most helpful in assembling complicated objects when a definite sequence of assembly must take place. Another typical use is in appliance and cabinetry service manuals.

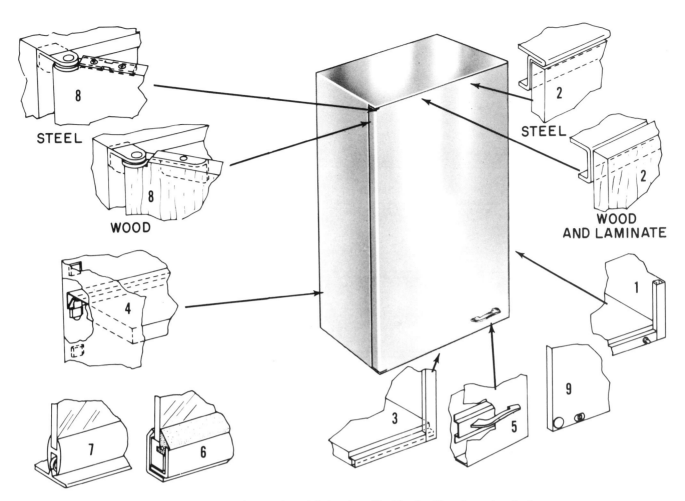

STEEL

STEEL

WOOD

WOOD AND LAMINATE

Fig. 6-17. An exploded pictorial drawing. (St. Charles Manufacturing Co.)

OPEN SHELF WALL UNIT

Make a parallel perspective sketch of the open shelf wall unit in the space below. Do not dimension.

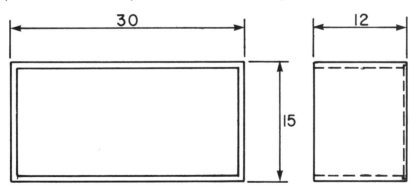

Sketching Assignment 6−6
OPEN SHELF WALL UNIT

Make an angular perspective sketch of the open shelf wall unit in the space below. Dimension the sketch.

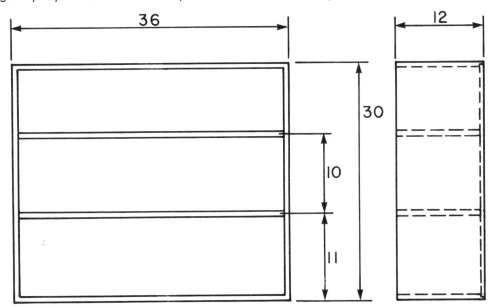

Unit 7
Orthographic Projection Drawings

A set of working drawings along with written specifications make up the "language of construction." Anyone working in the building field must be able to read these drawings and specifications. "Reading" construction drawings involves: understanding the various drawings used in a set of prints; knowing their relationship to each other; identifying the symbols and notes placed on the drawings. This chapter will assist you in developing a logical approach to the study of construction drawings and the relationship of one view to another.

Orthographic Projection

The series of views which make up a set of architectural drawings are all related to each other by a system known as ORTHOGRAPHIC (ortho-graf-ick) PROJECTION. The different views in orthographic projection are arranged in a systematic way so the user can form a mental picture of the object to be built, Fig. 7-1.

The architect or drafter separates the different views of a building into floor plans, elevations,

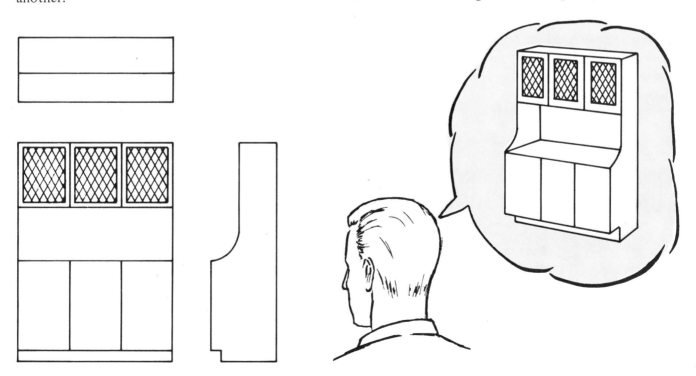

Fig. 7-1. Forming a mental picture of an object from a construction drawing.

sections and details so that its size, shape and construction may be clearly shown. Due to the size of some building, only one view may appear on a sheet. The skilled construction worker reading the print must be able to visualize (form a mental picture of) the building as a whole. Understanding how these views are developed to form a set of prints will help you in the visualization process.

Projecting the Views

The views of an orthographic drawing are projected at right angles (90 deg.) to each other and have a definite relationship. The best way to visualize this is by cutting and unfolding a cardboard box, as shown in Fig. 7-2.

The front view has remained in position. The four adjoining views have revolved on their "hinges" 90 deg., with the front view bringing them into the same plane, Fig. 7-2. The top view is above, the right side to the right, the bottom view is below and the left side is to the left. The rear view is shown to the left of the left side view, but it could be shown in alternate positions.

In architectural drawings, the different views of the building (floor plans and elevations) are obtained in the same manner. Imagine the building was enclosed in a large glass box, Fig. 7-3(a). Each view is projected toward its "viewing plane," Fig. 7-3(b), then unfolded and brought into plane with the front view, Fig. 7-3(c).

Due to the size of most architectural structures, the different views usually are separated and placed on individual sheets. Blueprints are made from these separate drawings and fastened together to form the set of prints for a particular job. The bottom view usually is not shown and the top or "roof plan" is shown only for complex structures.

Roof Plan View

The plan view is the view of the building from directly above. This would be the "roof plan" which is used only when the roof structure is complex and a drawing is needed to show its layout. For less complicated structures, the roof plan is omitted and the roof construction is adequately shown in the various elevation views and any necessary detail views.

Various types of roof construction are illustrated in Fig. 7-4. The roof slope or pitch usually is shown as a ratio of rise to span:

$$\text{Pitch} = \frac{\text{Rise}}{\text{Span}} \quad \text{or}$$

This roof ratio means for every 12 units of run (horizontal measurement), there are 5 units of rise (vertical measurement).

The construction details of eaves (roof overhang) usually are shown in wall sections or an elevation.

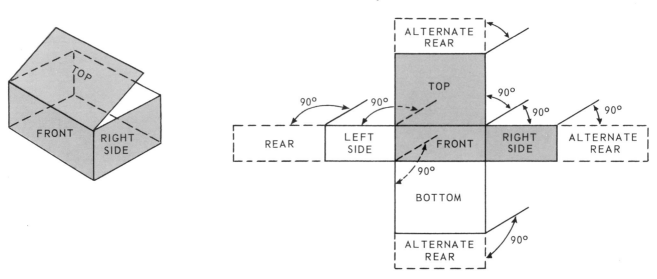

Fig. 7-2. Projection of orthographic views shown by unfolded box.

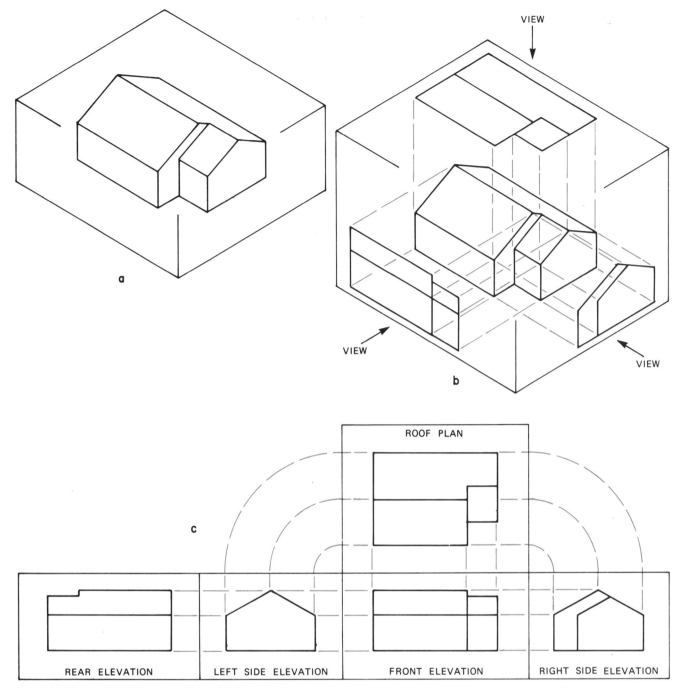

ROOF PLAN

REAR ELEVATION   LEFT SIDE ELEVATION   FRONT ELEVATION   RIGHT SIDE ELEVATION

Fig. 7-3. "Glass box" used to project views in orthographic projection.

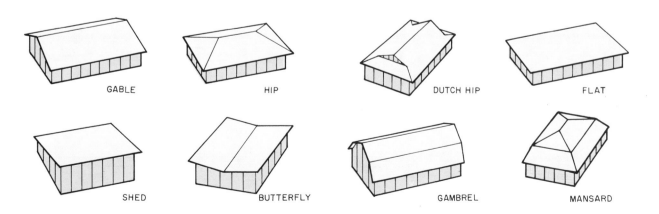

GABLE   HIP   DUTCH HIP   FLAT

SHED   BUTTERFLY   GAMBREL   MANSARD

Fig. 7-4. Types of roof construction.

Floor Plan

The floor plan is possibly the most important of all drawings. It is regarded as the "key" sheet in a set. By reviewing the floor plans and relating these to the elevation drawings, you can visualize what the structure will look like when built, Fig. 7-5.

Like the roof plan, the floor plan is a view from above. It represents a horizontal section at about eye level on each floor. The floor plan includes features such as doors, windows, cabinets (upper and lower), medicine chests and stairways. However, features which fall above or below this viewing plane (lower level of a fireplace, etc.) are included

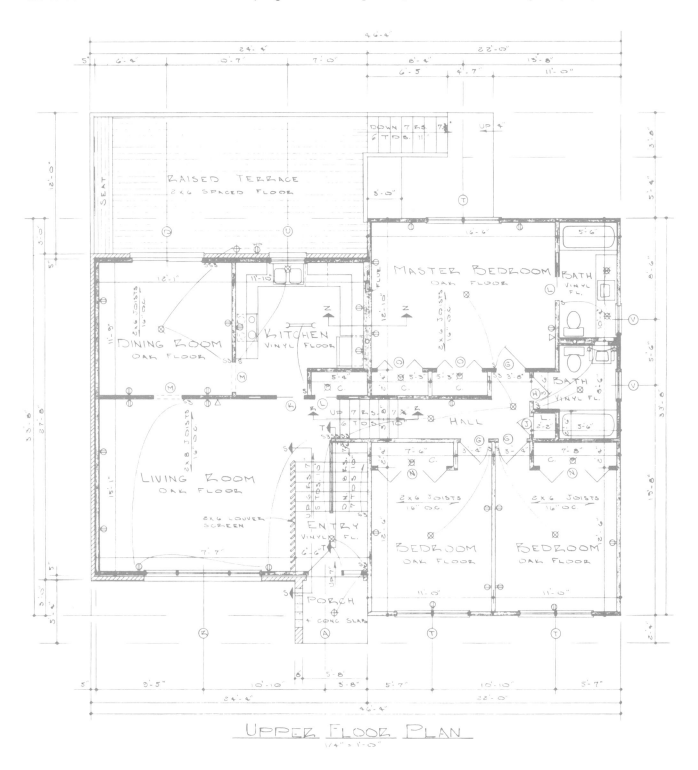

Fig. 7-5. The floor plan of a building showing the upper floor of two floors. (Garlinghouse Plan Service)

to give a complete accounting of all features. Usually, detail or sectional views of fireplaces and special cabinetry are drawn to clarify their construction.

Your study of the floor plan should start with a review of the general layout. Get an idea of the room arrangements, halls and storage before studying details of construction. This will help you to understand the complete set of plans.

Floor plans are drawn to an exact scale (usually 1/48 size: 1/4" = 1'-0"). However, you should rely on the dimensions shown and not scale the drawing. See Fig. 7-5. A separate drawing is made for each floor, including the basement. Note that the load from one floor is transferred to the floor below by supporting members or partitions.

Features on the floor plan may be referenced to other drawings in the set to further clarify the construction.

Foundation Plan

The foundation plan is similar to the floor plan. It shows the foundation wall and footings. If a basement is a part of the structure, it usually is shown on the foundation plan, Fig. 7-6. All wall openings, stairs, chimneys, etc., are shown.

Elevations

Elevations, Fig. 7-7, are exterior views of a building as seen by a person looking at each side. Elevation drawings show features such as the style of the building, doors, windows, chimneys and moldings. Any feature on an elevation drawing that does not have sufficient clarity will be shown on a larger scale in a detail drawing.

Elevations are designated as Front, Right Side, Left Side and Rear. They also may be identified by the direction the elevation faces. When the building faces EAST, the front elevation would become the EAST ELEVATION, and so on for the SOUTH, WEST and NORTH ELEVATIONS.

Interior elevation drawings also may be provided to show the type and construction of a particular interior wall or area.

Because the building is designed with the inside space in mind, the floor plan usually is drawn prior to the elevations. Then, linear measurements are projected or taken from the floor plan. Vertical or height measurements come from specifications or standards for the particular type of building. Normally, floor heights, type of doors, windows and other features are indicated on the elevations. Symbols may be used on elevation drawings to indicate type of exterior finish or material.

Below grade (ground level) features of the building, such as basement or foundation walls and footings, are shown in hidden lines on elevations.

Sections

The purpose of a set of blueprints is to show the construction details of a building. Plan and elevation views give most of the information needed. Sometimes, however, it is necessary to show the "inside" of a wall, cabinet or roof structure to clarify construction procedures. When the drawing is an imaginary "cut" through a wall or other feature, it is known as a SECTIONAL VIEW. See Fig. 7-8.

Sections may be provided for walls, cabinets, chimneys, stairs and any other feature whose construction is not shown on the plan or elevation views. Sectional views show how various components and materials are assembled to accomplish the desired result.

There are two other useful types of sections. These are the sections that "cut" completely through a structure to show more clearly the overall construction procedures. Sections which pass through the long dimension of a building are called LONGITUDINAL sections. A TRANSVERSE section is a "cut" through the short dimension of a building.

Details

Due to the scale at which construction drawings are usually made, certain features are not clearly shown on the plan, elevation or sectional views. These features usually require a larger scale illustration to provide information necessary for construction. For example, see cornice drawn in Fig. 7-9.

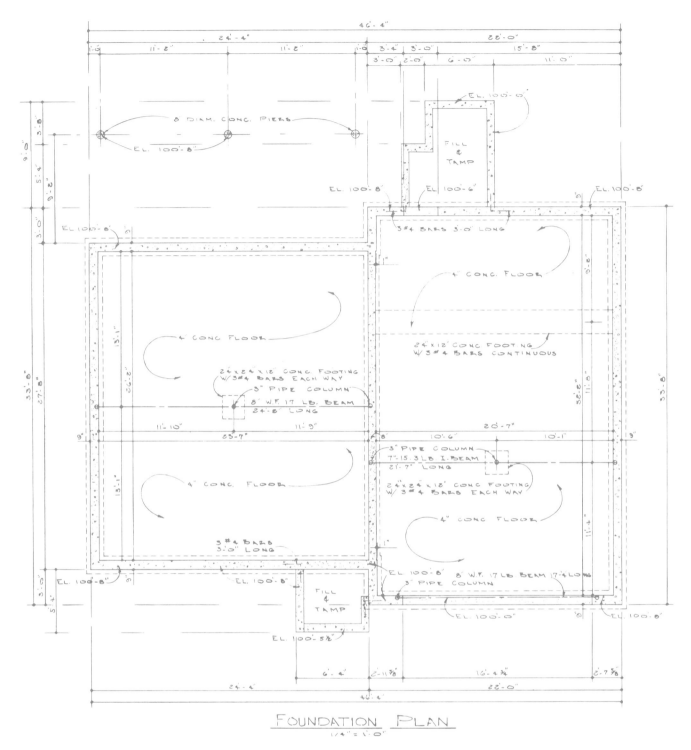

FOUNDATION PLAN

1/4" = 1'-0"

NOTE:
FOUNDATION WALLS UNDER BRICK VENEER WALLS
ARE 9" THICK WITH 18x8" CONC. FOOTINGS.
FOUNDATION WALL UNDER FRAME WALL IS 8"
THICK WITH 16x8" CONC. FOOTING.
FOUNDATION WALLS UNDER PORCH ARE 8" THICK
WITH NO FOOTINGS.

Fig. 7-6. A foundation plan of a building.   (Garlinghouse Plan Service)

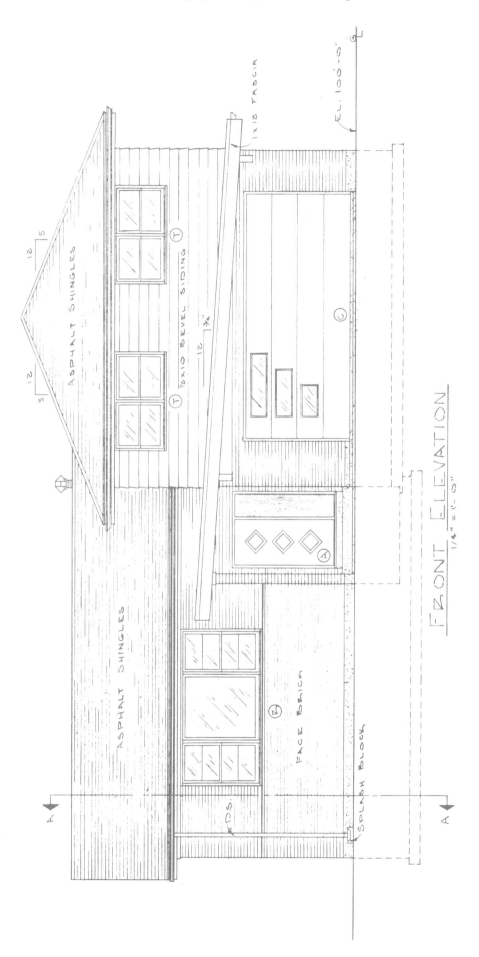

Fig. 7-7. A front elevation of the building shown in Fig. 7-5. (Garlinghouse Plan Service)

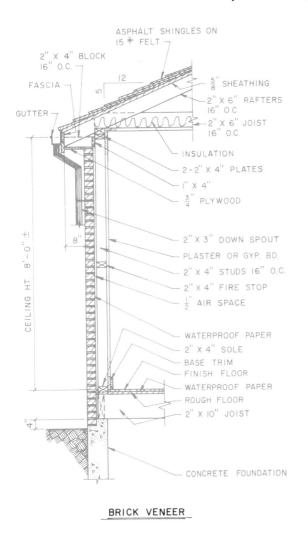

BRICK VENEER

Fig. 7-8. A sectional view shows construction details not in plan view and elevation view.

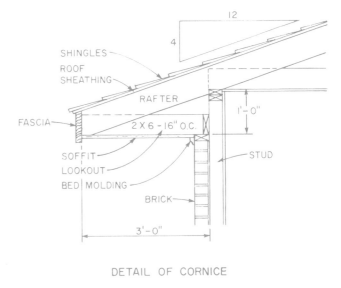

DETAIL OF CORNICE

Fig. 7-9. A detail drawing of a cornice.

Detail drawings may be placed on the same sheets as the plan or elevation views where the detail is shown in the building. Otherwise, the detail drawings may be found on separate sheets and referenced.

Set of Construction Blueprints

Small construction jobs usually include all necessary information such as style, structural, plumbing and electrical on the same plan and elevation drawings. Larger construction projects are more complicated and, rather than crowd the drawings with too much information, the set of plans may be divided into groups according to types of construction:

A - Architectural Layout and Design (plot plan, elevations, framing and building details).

S - Structural (wood, concrete and steel of the superstructure)

M - Mechanical (plumbing, heating, ventilation and cooling)

E - Electrical (power and lighting systems)

Some architects and engineers further divide the set of blueprints by adding:

U - Utility Site Plan (showing public or municipal electrical and plumbing supply lines)

P - Plumbing (waste and water supply systems)

H - Heating, Ventilating and Air Conditioning Systems.

Other architects simply number the sheets of a set of blueprints and use no letter classification.

Title Block

The Title Block on architectural drawings includes the architectural business name and address, date, plan number and sheet number. The amount of additional information included in the title block varys with the architect. Some include the project name and address, revision block and architect's seal of registry.

Orthographic Projection Drawings

Study the pictorial views and match each orthographic drawing with its pictorial drawing by inserting the correct letter in the space provided.

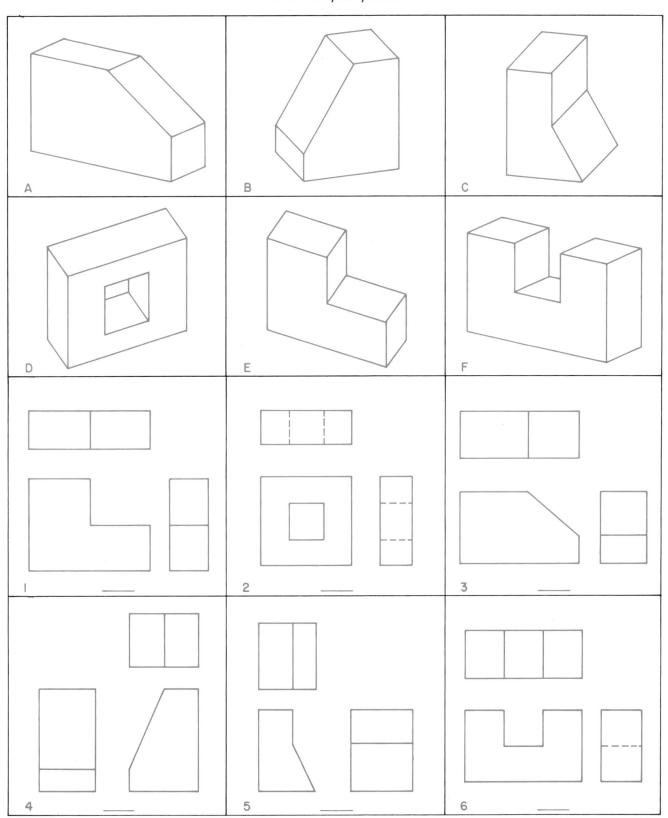

Blueprint Reading Activity 7-2
IDENTIFICATION

Use this architectural rendering of a split level house to answer questions on page 63. (Continued)

Use this floor plan of the house shown on page 62, to answer the following questions.

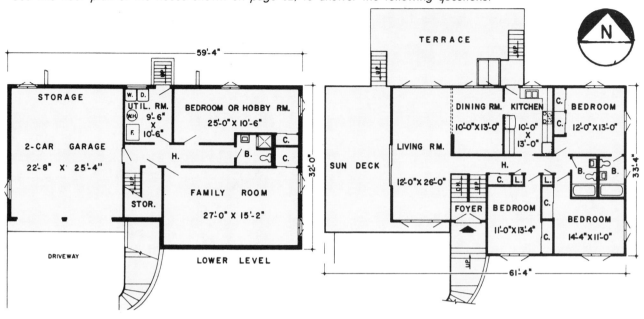

1. The front entrance door will be shown in the _____ elevation.   1. _____

2. What is the style of the roof?   2. _____

3. How many windows will be shown in the South elevation? In the North elevation?   3. _____
                                                                                                                                      _____

4. How many closets are in the northeast bedroom?   4. _____

5. The dining room has a wide opening into the living room and a smaller opening into the _____.   5. _____

6. There are _____ full baths on the upper level.   6. _____

7. How many closets are in the hall on the upper level?   7. _____

8. The exterior door of the living room leads to the _____.   8. _____

9. The garage is on the _____ level.   9. _____

10. The garage doors will appear in the _____ elevation.   10. _____

11. How many windows are in the garage?   11. _____

12. An exterior stairway on the lower level permits direct access to the outdoors from the _____.   12. _____

13. Access to the bath on the lower level is from the _____.   13. _____

14. Space beneath the foyer is utilized for _____.   14. _____

15. The East elevation will show _____ windows.   15. _____

Blueprint Reading Activity 7 — 3
SKETCHING ORTHOGRAPHIC VIEWS

Sketch the plan view and the two elevation views shown in each pictorial. The views should be correctly positioned with each other.

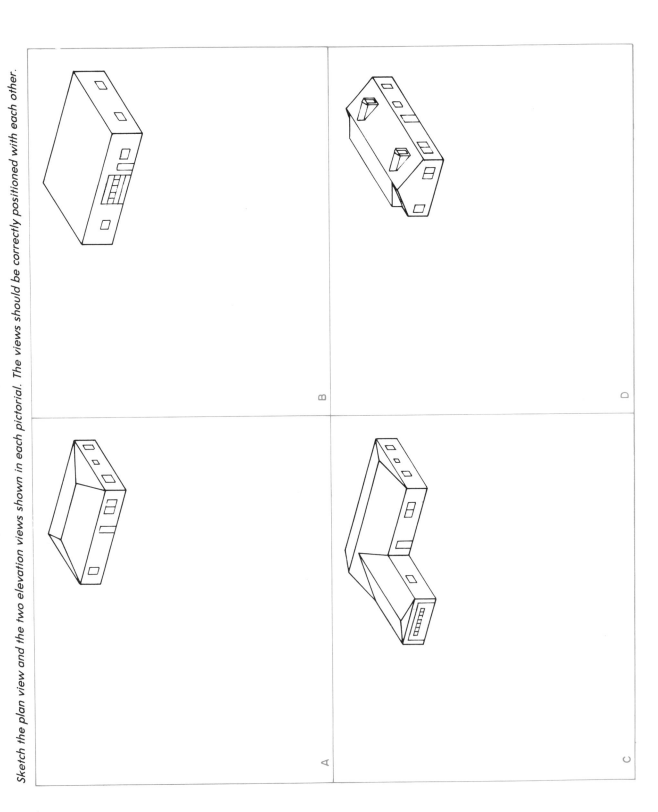

A

B

C

D

Unit 8
Construction Dimensioning Techniques

Dimensions are an essential part of blueprints. Therefore, the person reading a print must understand the various dimensioning techniques so that correct readings may be made. You will find that drafting standards vary greatly throughout the country even though a considerable effort is being made to unify procedures.

Standard procedures for dimensioning are discussed in this unit. A careful study of these procedures will enable you to properly interpret construction drawings. The few variations found in dimensioning techniques will be noted for your clarification.

A print made from a construction drawing which has been properly dimensioned by the architectural drafter requires no additional mathematical calculations to obtain needed dimensions.

Dimension Lines

The character of dimension and extension lines was described in Unit 4. Architects and drafters use various means of terminating dimension lines at extension lines, Fig. 8-1. The slim arrowheads are most common with other forms shown at b, c and d used to conserve time and give individuality to the work. Dimension figures usually are shown above the dimension line. See Fig. 8-1.

Scale of Drawings

Construction projects are drawn to a reduced scale to make the blueprints useful on the job. The scale of a particular floor plan, elevation or detail is indicated on the sheet, either in the title block or beneath the drawing itself. An example of scale indication is: SCALE - 1/4" = 1'-0".

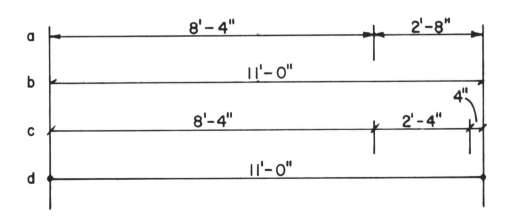

Fig. 8-1. Forms used on construction drawings to terminate dimension lines.

In addition to referring to the relative size a drawing has been drawn, the term "scale" also refers to the instrument (ruler) used in laying out measurements to scale.

The scale most commonly used for floor plans in the customary (English) measurement system is 1/4" = 1'-0" or 1/48 size (there are forty eight units of 1/4 inch in 12 inches). This is commonly referred to as the quarter scale. Scales for detail drawings range from 1/2" = 1'-0" to full size.

Scale Drawings - Metric Measurement System

The metric system of measurement has seen little use in the U.S. primarily because metric construction standards have not been established. Industry groups are at work developing standards. Once metric standards have been adopted and metric modular materials become available, metric dimensioning will be used extensively.

How to Read the Scale - Customary System

As a construction worker, you will be reading dimensions given on a print and laying off measurements on the job. The necessary dimensions should be provided on the print. However, there may be times when a particular dimension is desired but has been omitted or not considered necessary. Also, you may be required to sketch a detail, or lay it out to scale. Then, it would be necessary to read a scale (instrument).

The following steps will assist you in reading the customary system quarter scale:
1. Locate the scale marked 1/4, Fig. 8-2.
2. Starting at zero, the scaled measurements of feet are numbered to the left, such as 2, 4, 6, 8 and so on.
3. Inches are marked in the section to the right of zero.
4. To lay off a measurement of 6'-3", start with

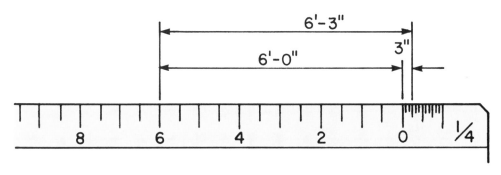

Fig. 8-2. Laying off a measurement on architect's scale.

The metric scale most closely representing the customary system quarter scale (1/48 size) is the 1:50 scale (1/50 size) in which two centimeters on the drawing equals one meter (100 cm) on the actual object. All metric scales are size scales.

the line indicating 6" and move to the zero, and on to 3". See Fig. 8-2.
5. A measurement of 11'-8" would start at the appropriate line indicating 11 feet and extend past zero to 8 inches.

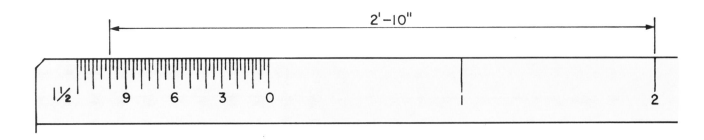

Fig. 8-3. A measurement on the 1 1/2" - 1'-0" scale.

Architectural details frequently are drawn to larger scales, such as 1 1/2'' = 1'-0''. See Fig. 8-3. This scale is read in the same manner as the quarter scale.

The Fractional Rule as a Scale

When an architect's scale is not available, the fractional rule may be used to take or lay off measurements. For a measurement on the 1/4'' = 1'-0'' scale, let each 1/4 inch on the rule represent one foot of actual measurement on the project. Fractional parts of the 1/4'' would represent the appropriate number of inches on the project. For

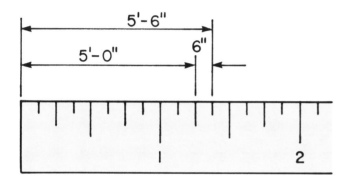

Fig. 8-4. Using the fractional rule to scale measurements on the 1/4'' = 1'-0'' scale.

example, a scaled reading of 5'-6'' would be five quarters of an inch for the 5 feet and one eighth of an inch for the 6 inches or 1 3/8 inches, Fig. 8-4.

It should be emphasized that a good drawing would include all necessary dimensions. Caution should be exercised in "scaling" a drawing for measurements not provided. All such measurements should be cross checked with other dimensions and verified by your supervisor.

How to Read the Scale - Metric System

Readings on the metric scale are made in the same manner as on the customary scale. The unit of metric linear measure in construction is the meter. A reading of 2.5 meters on the 1:50 metric scale is shown in Fig. 8-5. The unit of metric linear measure in cabinet work is the centimeter (cm). A measurement of 82 centimeters on the 1:20 metric scale is shown in Fig. 8-6.

Dimensioning Practices on Floor Plans

EXTERIOR WALLS of solid masonry construction are dimensioned to the exterior surface, Fig. 8-7. Dimensions for exterior walls of frame and brick-veneer buildings usually start at the outside

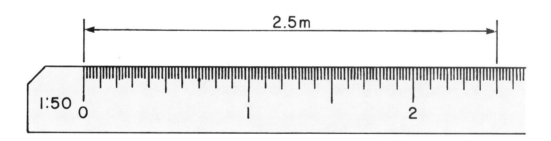

Fig. 8-5. A reading of 2.5 meters on the metric 50 scale.

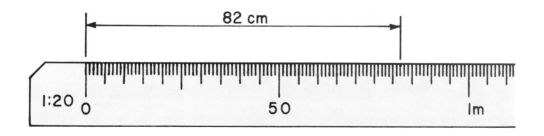

Fig. 8-6. A reading of 82 centimeters on the metric 20 scale.

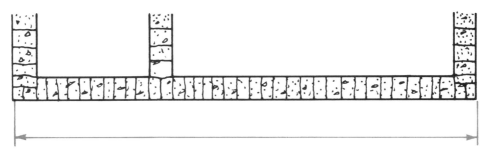

SOLID MASONRY EXTERIOR

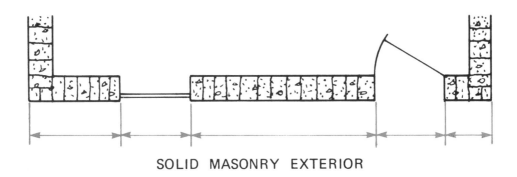

SOLID MASONRY EXTERIOR

Fig. 8-7. Dimensioning practices for solid masonry and frame walls.

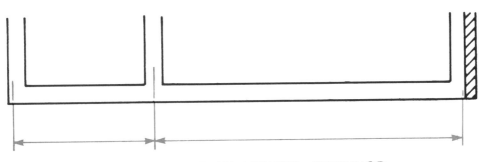

FRAME OR BRICK VENEER EXTERIOR

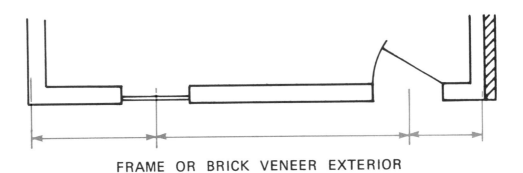

FRAME OR BRICK VENEER EXTERIOR

Fig. 8-8. Dimensioning windows in solid masonry and frame buildings.

surface of the stud wall, Fig. 8-8. Some architects show the dimension to the outside of the masonry on brick-veneer as well. This method provides the workers with the dimensions necessary for laying out the sole plates over subfloors and in locating window and door openings before the sheathing (structural covering over studs) is added.

Some architects dimension exterior walls of single-story frame construction by starting the dimensions at the surface of the sheathing. This surface should align with the foundation wall on the detail drawing. There is a way to clarify the dimensioning practice used on exterior walls. If the scale of the drawing is too small to show the dimension clearly, a note may be added as follows:

NOTE: EXTERIOR DIMENSIONS ARE TO OUTSIDE EDGE OF STUDS; INTERIOR DIMENSIONS ARE TO CENTER OF STUDS.

In the absence of a note or clarity of dimensions, you should calculate the exact location of the exterior dimension by referring to other prints of floor plans, elevations or details.

INTERIOR WALLS are usually dimensioned to the center of partitions, Fig. 8-8, however, some architects follow the practice of dimensioning to partition surfaces, then dimensioning the thickness of each partition.

WINDOW AND DOOR OPENINGS are located by dimensions to their center lines for frame construction, Fig. 8-8. For solid masonry construction, Fig. 8-7, these openings are dimensioned to the edges of the masonry surface openings. Doors or windows in narrow areas may not be dimensioned for location since it is obvious they would be centered in the space available.

Dimensioning Practices on Elevations

Dimensions provided on elevation drawings are those related to the vertical plane, since most horizontal dimensions are included on plan drawings. Such dimensions as footing thickness, depth of footing below grade, floor and ceiling heights, window and door heights and chimney height are provided on elevation drawings, Fig. 8-9. In addition to height dimensions, information is provided through notes on grade information, materials for exterior walls and roof, and special details.

Roof slope usually is given on a drawing as a slope triangle. The diagram represents the ratio between the rise and run (1/2 the entire span of the building). A typical slope would be 4:12 or 4 units of rise for each 12 in the run. See Fig. 8-9.

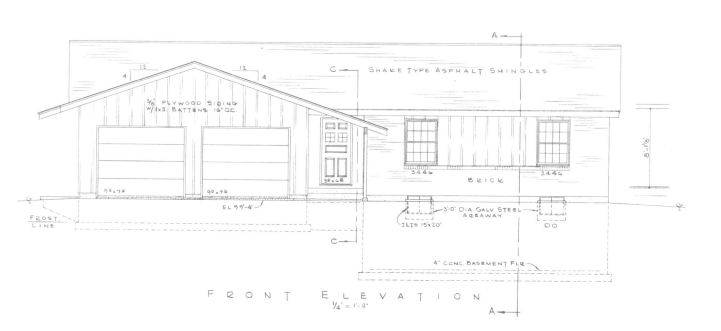

Fig. 8-9. Typical dimensions found on elevation drawings.

Dimensioning Practices on Sections and Details

For greater clarity, section and detail drawings frequently are drawn to a larger scale than plan and elevation drawings. Detail dimensions showing thicknesses of finished and sub-floor materials, joist sizes, molding location, etc., provide essential construction information. See Fig. 8-10.

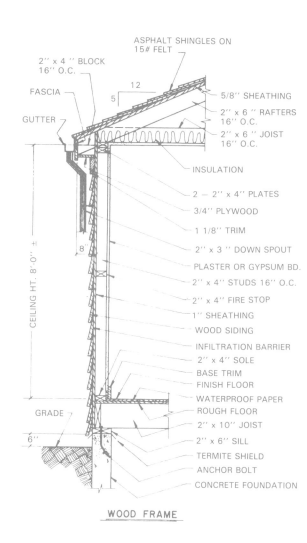

ASPHALT SHINGLES ON 15# FELT

2'' x 4 '' BLOCK 16'' O.C.

FASCIA

GUTTER

5/8'' SHEATHING

2'' x 6 '' RAFTERS 16'' O.C.

2'' x 6 '' JOIST 16'' O.C.

INSULATION

2 — 2'' x 4'' PLATES

3/4'' PLYWOOD

1 1/8'' TRIM

2'' x 3 '' DOWN SPOUT

PLASTER OR GYPSUM BD.

2'' x 4'' STUDS 16'' O.C.

2'' x 4'' FIRE STOP

1'' SHEATHING

WOOD SIDING

INFILTRATION BARRIER

2'' x 4'' SOLE

BASE TRIM

FINISH FLOOR

WATERPROOF PAPER

ROUGH FLOOR

2'' x 10'' JOIST

2'' x 6'' SILL

TERMITE SHIELD

ANCHOR BOLT

CONCRETE FOUNDATION

CEILING HT. 8'-0'' ±

GRADE

WOOD FRAME

Fig. 8-10. Section and detail drawings provide essential construction dimensions.

Blueprint Reading Activity 8—1
READING BLUEPRINTS FOR DIMENSIONS
FRAME RESIDENCE

Refer to Prints 8-1a and 8-1b in the Large Prints Folder to answer the following questions.

1. Give the scale of the floor plan.

1. _1/2" = 1'-0"_

2. What is the overall length and width of the house?

2. Length: _83'-0"_
   Width: _34'-0"_

3. Locate the Family Room-Kitchen wall from the rear exterior and explain where the dimensions start and end.

3. _19'-10" x 11'-10"_

4. Give the dimensions for the layout of the Living Room, and explain where the dimensions start and end.

4. _20'-0 x 13'-6'_

5. What are the layout dimensions for the Rear Bedroom?

5. _11'-6" x 12'-6"_

6. Give the layout size of the Hall Bath.

6. _7'-2" x 9'-0"_

7. What is the width of the Breezeway?

7. _11'-0"_

8. Locate the Hall Bath window from the nearest corner.

8. _____

9. How far is the center of the front door from the bedroom end of the house.

9. _____

10. What is the distance from the finished floor of the Basement to the bottom of the first floor joists?

10. _7'-0"_

11. Give the distance from the sub-floor to the bottom of the ceiling joists.

11. _8'-1"_

12. What is the distance from the exterior stud face to the rafter plumb cut?

12. _3'-0"_

13. How thick is the foundation wall?
    Basement floor slab?

13. _4" 8"_

2'2"
4'4"
2'0 x 9'0 x

14. Give the nominal size of the following structural members:
    a. Floor joists
    b. Wall studs
    c. Ceiling joists
    d. Rafters

14. a. _2″ x 8″_
    b. _2′ x 4_
    c. _2′ x 6″_
    d. _2″ x 6″_

15. What size and type of column is used to support the beam in the Basement?

15. _3½″ Round Pipe Column_

16. Give the size of the beam.

16. _~~3½″~~ W7″ x 15.3 x 5′_

17. What is the size of the foundation wall footing? Footing for the pipe column?

17. _____

18. What size drain tile is to be placed outside the footings?

18. _____

19. What is the size of the beam over the Foyer that supports the ceiling joists?

19. _____

20. How are the ceiling joists fastened to the beam over the Foyer?

20. _____

Construction Dimensioning Techniques

Refer to Prints 8-2a and 8-2b in the Large Prints Folder to answer the following questions.

1. What is the scale of the foundation and floor plan drawings?

1. *Foundation ½" = 1'-0" and Floor ⅛" = 1'-0"*

2. Give the overall length and width dimensions of the foundation.

2. *36'-0"*

3. Locate the beam support column at the head of the stairs from the interior side of the basement fireplace and front walls.

3. _____

4. How thick are the exterior foundation walls? Interior walls?

4. *9" exterior and 8" interior*

5. Give the cross section size of the footings under the exterior foundation wall.

5. _____

6. How far below the top of the foundation wall is the recess for the brick?

6. _____

7. Give the following dimensions for the garage:
   a. Width from exterior side of wall to opposite exterior side of wall.
   b. Foundation opening for garage door.
   c. Greatest length inside foundation wall back to front.
   d. Distance across the width from exterior stud face to exterior stud face.

7. a. _____
   b. _____
   c. _____
   d. _____

8. Give the overall length and width of the frame wall of the house from exterior stud faces.

8. Length: _____
   Width: _____

9. Locate the center of the front door opening from the nearest foundation wall.

9. _____

10. What is the interior size of the living room?

10. _____

11. Locate the wall between the kitchen and dining room from the den end of the house.

11. _____

12. Give the following distances:
    a. Basement floor to underside of joists.
    b. Subfloor to second floor joists.
    c. Second floor subfloor to ceiling joists.

12. a. _____
    b. _____
    c. _____

13. What is the roof slope?

13. _____

14. With reference to the basement stairs, give the number and dimensions for the following:
    a. Risers
    b. Treads

14. a. _____
    b. _____

15. What floor covering is specified for the kitchen and utility room?

15. _____

PART 3
CONSTRUCTION PROCEDURES BLUEPRINTS

Unit 9
Construction Materials

Construction workers will find frequent references to materials on a set of blueprints. Most of these materials are carefullly detailed in the specifications, and this source should be relied upon to specifically identify the material concerned.

The purpose of this unit is to provide you with an overview of the most common materials used in construction. Also discussed, where appropriate, are the symbols used on construction drawings. Symbols are general classifications useful in reading a drawing to better understand the construction job.

Concrete

CONCRETE is one of the oldest building materials, having been used by the Romans as early as 100 B.C. Concrete is a mixture of cement, sand and aggregate (usually gravel) with water. When first mixed, it is in a plastic state and can be cast in place to take the shape of the form. The concrete begins to set (harden) after a few minutes if left undisturbed.

Hardening of the concrete is not caused by drying but by chemical reaction in the cement. Most mixtures of concrete set within 12 to 24 hours, depending on the atmospheric temperature, the volume of the pour and certain other factors, such as admixtures used to speed or retard setting. When temperatures are below 70 deg. F (21.1 C), chemical reaction slows. Very little chemical reaction takes place at 30 deg. F (-1.1 C) or below. Concrete continues to harden for months or even years, but most pours reach their load bearing or design strength within 28 days. Forms may be removed after several days, depending on the volume of the pour.

Types of Cement

CEMENT is the material which binds the concrete mix together in its solid form. There are many types of cement but the most common type used for general construction is called normal portland cement. Another type used in construction is white portland cement. It is light-colored and used chiefly for architectural effects. White portland cement is made from carefully selected raw materials and develops the same strength as the grayish, normal portland cement.

Other types of cements, aggregates and admixtures are available to produce special types of concrete. Some have qualities of freeze resistance in setting in cold climates. Some have low-heat generation for construction of large projects such as dams. Others have high-early strength to produce concrete which sets faster than normal, permitting earlier form removal and thus speeding construction. Still others are more resistant to deterioration of sulfates and alkalies in the soil.

Concrete Mixes

Basically, a CONCRETE MIX should be designed to produce the desired result. A 1:3:4 mix is a concrete that contains, by volume, one part cement, three parts sand and four parts aggregate (gravel). Enough water is added to the mixture to cause the needed chemical reaction and to make it plastic enough to flow into the forms. Too much water can reduce the ultimate strength of the concrete.

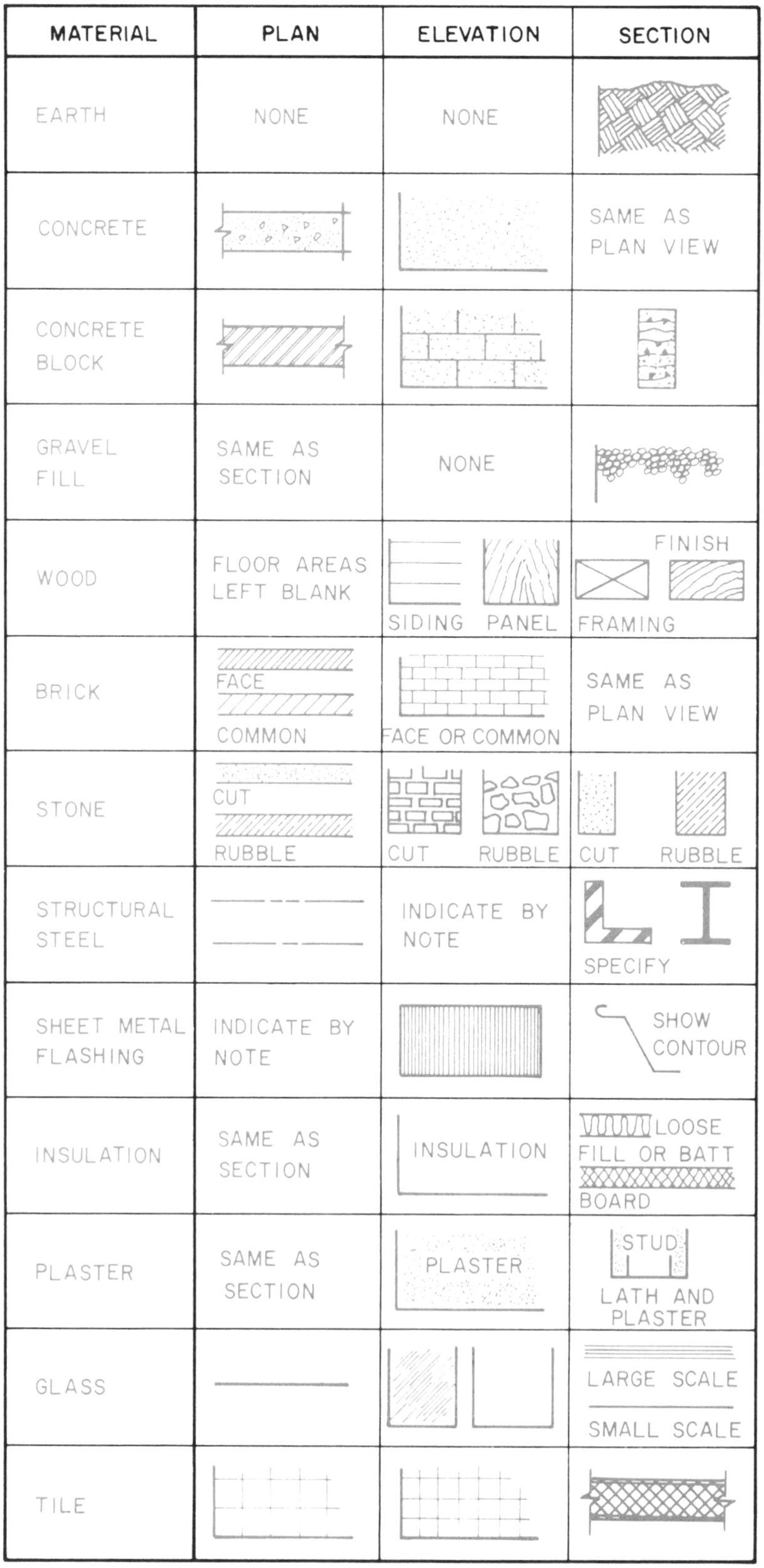

MATERIAL	PLAN	ELEVATION	SECTION
EARTH	NONE	NONE	
CONCRETE			SAME AS PLAN VIEW
CONCRETE BLOCK			
GRAVEL FILL	SAME AS SECTION	NONE	
WOOD	FLOOR AREAS LEFT BLANK	SIDING PANEL	FINISH FRAMING
BRICK	FACE COMMON	FACE OR COMMON	SAME AS PLAN VIEW
STONE	CUT RUBBLE	CUT RUBBLE	CUT RUBBLE
STRUCTURAL STEEL		INDICATE BY NOTE	SPECIFY
SHEET METAL FLASHING	INDICATE BY NOTE		SHOW CONTOUR
INSULATION	SAME AS SECTION	INSULATION	LOOSE FILL OR BATT BOARD
PLASTER	SAME AS SECTION	PLASTER	STUD LATH AND PLASTER
GLASS			LARGE SCALE SMALL SCALE
TILE			

Fig. 9-1. Symbols used for construction materials.

Any material added to the concrete mix other than cement, sand, aggregate and water is known as an ADMIXTURE. Admixtures are used to make the mix more workable, retard or speed hardening, freeze resistant or other desired quality.

Symbols

Concrete is represented on the drawing in the section view as small dots with a scattering of small triangular shapes, Fig. 9-1. Concrete on plan views, such as the foundation plan, is shown blank or, on some drawings, as small portions of the same symbol as on section views. If symbols for concrete are used on elevation views, it is small dots. If the concrete area is sizable on elevation drawings, the symbol is used only on small proportions.

Reinforced Concrete

Concrete has great compression strength, but very little tensile (pulling) strength. To overcome this weakness in tensile strength, concrete is cast with steel bars which have great tensile strength. As the concrete hardens it grips the steel to form a bond. Reinforcing bars are usually round with projections formed in the rolling process to add strength to the bond in reinforced concrete. (The size designation of reinforced-steel bars is discussed later in this chapter.) Sheets of wire mesh are used to reinforce concrete slabs where heavy traffic loads are anticipated, Fig. 9-2.

When concrete has internal stresses to counteract stresses caused by external loading, it is known as PRESTRESSED CONCRETE. The effect of a prestressed concrete beam can be illustrated by a person picking up a number of individual wooden blocks, Fig. 9-3. The pressure placed on the ends of the row of blocks keeps the blocks from collapsing in the middle and falling. The row of blocks under pressure form a beam. The harder the pressure, the stronger the beam. The prestressed concrete beam acts on the same principle.

Concrete is prestressed by casting it around steel cables or rods which are placed under tension (pulled) to as much as 250,000 psi. When the concrete hardens, tension on the steel causes a stressed condition on the concrete. This gives it

Fig. 9-2. Welded wire reinforcement being used to strengthen a concrete floor. (Wire Reinforcement Institute)

greater strength in resisting load bearings. Concrete slabs also may be prestressed to provide greater strength. With the development of prestressed concrete, structural members may be designed to use half the concrete and one-fourth the steel required for conventional reinforced concrete members.

Fig. 9-3. The principle of strength resulting from pressure placed on prestressed concrete.

77

Beams or slabs in which the steel has been tensioned before casting the cement are known as PRETENSIONED CONCRETE. The tension on the steel can also be done after casting the concrete, providing the steel rod or cables have been encased in light metal or paper tubes so that the concrete does not bond to the cable. After the concrete has hardened, the steel is placed under tension. This is known as POSTTENSIONED CONCRETE.

Masonry

Masonry products are natural or manufactured building units such as stone, brick, concrete, tile, glass or gypsum. Masonry construction is the completed project made of stone, brick, concrete, etc., or combination thereof, bonded with mortar.

Stone Masonry

The most common stones used in STONE MASONRY construction are granite, limestone, marble, sandstone and slate. Like concrete, stone has been used as a building material for many centuries. In the past, stones were used for structural, roofing and as finished material. With the development of new materials and methods of construction, and because of the cost of stones, they are used today largely for their decorative value.

Most stones are removed from quarries and sent to a finishing mill for final dressing. Some stones are used in their original shapes and surface finishes. Other are purchased for construction jobs that require stones of a certain shape, size and finish. These stones are known as CUT STONES.

Brick Masonry

BRICK MASONRY, in contrast to stone masonry, uses manufactured products as its basic material units. Building brick are classified as:
ADOBE: Natural sun-dried clays or earth and a binder.
KILN BURNED: Natural clays or shales (sometimes other materials are added, such as chemicals to produce a certain color) molded to shape, dryed and fired for hardness.

SAND-LIME: Composed of a mixture of sand and lime, molded and hardened under steam pressure and heat.
CONCRETE: A mixture of portland cement and aggregates, molded into solid or cored units and hardened chemically.

There are many types and sizes of brick. Those most generally used in construction are building brick and face brick. Some special types are used to a lesser extent.

BUILDING BRICK, known as common brick, is the brick most often used in construction. It is used for walls, backing and other applications where appearance (color, shape and texture) is not of prime importance.

FACE BRICK is manufactured under more controlled conditions to produce bricks of certain dimensions, color and structural qualities. These bricks are more expensive than building bricks because of the care going into their manufacture. Face bricks with defects are often sold as common brick.

Special types of bricks include:
GLAZED BRICK for use in decorative and special service applications.
FIRE BRICK for use where masonry units are subjected to extreme heat, such as fireplaces, incinerators and industrial smelting furnaces.
PAVING BRICK for use in drives or area ways where abrasion is a factor.

Special bricks also are available in unusual shapes for window sills, rounded corners and special custom shapes required for a specific job where a special effect is desired.

Mortar

Mortar for stone and brick masonry usually consists of cement, hydrated lime, sand and water. The purpose of the mortar is to cement the units together as a whole and form a strong, durable bond.

Stone work usually requires a special type of mortar consisting of white portland cement, hydrated lime and sand to prevent stain caused by ordinary cements. Mortar in the joints usually is raked back from the surface as the stone is set. Later, the joint is pointed, using the same mortar or a colored synthetic sealant.

Trade associations recommend proportions of cement, hydrated lime and sand for best mortar results. Local building codes also set allowable limits. A typical mixture would be 1 part portland cement by volume, 1/4 hydrated lime and 3 parts sand. Enough water is added to make the mortar workable. Also, many brick masons use a special masonry cement containing plasticizing agents that make the mortar easily workable and waterproof when set.

Structural Clay Tile

Structural CLAY TILE is a product made of materials that are similar to those used in making brick, but it is a larger building unit than brick. It has many uses in construction, such as nonbearing partition walls, load-bearing walls, backup for curtain walls and as fireproofing around structural steel. Rectangular open cells pass through each unit, and the tile come in a variety of structural and design shapes and sizes.

Clay tile has largely been replaced with the hollow brick and concrete masonry units. Some typical hollow brick are shown in Fig. 9-4. Most masonry walls today are composite walls of a finish surface material and a less expensive backup material.

TERRA COTTA is a type of structural clay tile principally used for nonbearing ornamental and decorative effects in construction.

Concrete Masonry Units (CMU)

Another popular and widely used building material is the CONCRETE BLOCK, which is made from portland cement and sand, gravel or other

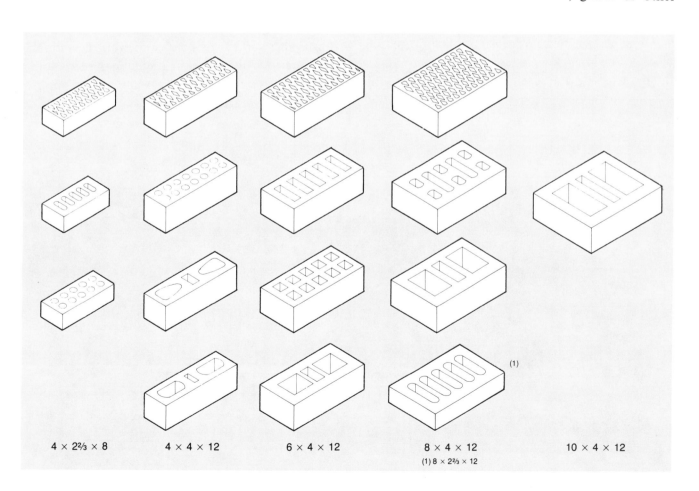

4 × 2⅔ × 8   4 × 4 × 12   6 × 4 × 12   8 × 4 × 12   10 × 4 × 12

(1) 8 × 2⅔ × 12

Fig. 9-4. Some examples of hollow brick design shapes. (Brick Institute of America)

aggregates. They are made in a variety of sizes, shapes and density to meet specific construction needs. The standard block is 7 5/8 x 7 5/8 x 15 5/8 inches, which lays up in a 8 x 8 x 16 inch module. Another common size is the 3 5/8 x 7 5/8 x 15 5/8 inch block. By using different type aggregates such as sand and gravel, expanded shale or pumice, their weight, load bearing and acoustical qualities can be controlled.

Concrete blocks with colored surfaces and special design features, and blocks known as "slump" block giving the appearance of rough adobe brick, Fig. 9-5, are available in some areas.

Gypsum Blocks

Gypsum masonry blocks are used primarily for interior, non-load-bearing wall partitions and fire resistant partitions and enclosures around structural steel. Made from gypsum and a binder of vegetable fiber, asbestos or wood chips, they may be given a plaster finish coat. Gypsum blocks have a face size of 12 x 30 inches and come in thicknesses of 2, 3, 4 and 6 inches. There is no special symbol for gypsum block, other than for plaster, Fig. 9-1, so a note on the drawing and the specifications would detail this building material.

Gypsum Wallboard and Lath

Gypsum wallboard and lath products consist of a core of air-entrained gypsum between two layers of treated paper. Wallboard comes in 4 x 8 foot sheets and longer. It is fastened to wood or metal studs with nails or screws. It varies in thickness from 1/4 to 1 inch, and the joints are sealed with a joint compound and paper tape to provide a wall with a smooth even surface. The wall may be painted, papered or given a surface texture to enhance its appearance.

Gypsum lath is available in 16 x 48 inch sheets with thicknesses of 3/8 and 1/2 inch. The lath is fastened to the studs and a three coat plaster process – scratch coat, brown coat and finish coat – is applied.

Glass Block

GLASS BLOCKS are made from two sections of glass, fused together, creating a partial vacuum which has good insulating qualities. They have a rough edge surface to bond well with the mortar. They are used in nonbearing situations such as interior walls, screens, curtain walls and windows. They are manufactured in three nominal sizes: 6 x

Fig. 9-5. A popular type of concrete masonry unit is the "slump" block.   (National Concrete Masonry Association)

6, 8 x 8 and 12 x 12 inches, with special blocks for turning a corner or forming a curved panel.

A note on the drawing identifies the glass block and the specifications detail the type and size.

Wood Products

WOOD continues to be one of the chief building materials in use today. It is used for structural framing (rough carpentry), trim, floors, walls and cabinetry (finish carpentry) and for many decorative effects in building. While some inroads have been made in substituting other materials in place of wood, it remains a valuable and most widely used residential construction material.

When wood is cut into pieces of uniform thickness, width and length, it is called LUMBER. Some of the lumber products are:
Rough framing members (usually 2 inches or more in thickness) including beams, headers and posts.
Finished lumber such as flooring, door and window trim, paneling and moldings.
Speciality items such as decorative and carved panels and doors, ornamental overlay designs and turned balusters (stair rail posts).

Kinds of Wood

Woods are broadly classified as either hardwoods or softwoods. Some of the more common types are:

HARDWOODS	SOFTWOODS
Maple	Ponderosa Pine
Oak	Southern Pine
Walnut	White Pine
Mahogany	Douglas Fir
Cherry	Hemlock
Birch	White Fir
Beech	Spruce
Elm	Eastern Red Cedar
White Ash	Western Red Cedar
Ash	Redwood
Gum	Cypress
Hickory	Yellow Poplar

This classification is not an exact measure of hardness or softness (because this varies), but rather is a general classification based on type of trees. In addition to hardness or softness, woods vary in strength, weight, texture, workability and cost. Building specifications usually indicate in detail the type and grade of lumber to be used in different parts of the construction.

PLYWOOD is a wood product consisting of several layers of lumber with the grain at right angles in each successive layer. An odd number of layers (3, 5, 7, etc., so the grain of the face and back are running in the same direction) are bonded together with an adhesive. The panels are finished to thicknesses of 1/8 to over 1 inch, usually 4 x 8 feet in size.

Interior plywood is bonded with an adhesive that is water resistant (will stand an occasional wetting). It is used in cabinetry, rough flooring and for finished walls. Exterior or structural plywood is bonded with a waterproof adhesive. It is used for exterior wall sheathing and finished walls, roof sheathing and concrete forms.

Because of its modular size and uniformity, plywood speeds construction and is considered an economical building material.

SYMBOLS: For a wood frame wall on the plan view, the usual practice is to leave the wall blank, Fig. 9-6(a). Some architects shade this area lightly so as to better outline the building and its partitions, Fig. 9-6(b). Section views that cut across structural framing members show these pieces with an "X" within each member, Fig. 9-1. Finished lumber (trim, facia boards, moldings) in section shows the wood end grain, Fig. 9-1.

Plywood, when shown in small scale, is represented with the same symbol as lumber. When in section and the scale permits, lines may be drawn to indicate the plies (not necessarily the exact number). In elevation views, wood siding and panels are represented as shown in Fig. 9-1. Grades and types of wood products are also detailed in the building specifications.

Glued-Laminated Timber

The process of LAMINATING (lumber bonded together in layers with an adhesive) wood beams,

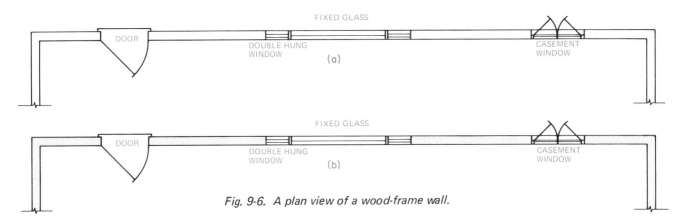

Fig. 9-6. A plan view of a wood-frame wall.

arches and other structural members has made it possible to span larger distances and to change traditional construction techniques. Wood members of nearly any size and shape are either made or can be fabricated for a special job, Fig. 9-7. These laminated products are made of kiln dried lumber and are prepared either for interior or exterior use depending on the type of glue used. These beams usually are prefinished at the factory and delivered to the job with a protective wrapping.

Metal Products

Many uses are made of metal in the construction industry. Large commercial buildings use structural steel by itself and in combination with concrete. All types and sizes of construction jobs make use of metal windows, doors, studs, beams, joists, wall facings, roofing, plumbing, electrical and hardware of various kinds. Without the use of metal, modern building would not be as extensive as it is today.

STRUCTURAL STEEL is the term applied to hot-rolled steel sections, shapes and plates 1/8 inch thick or greater. This includes bolts, rivets, bracings or devices used to complete the steel frame.

Structural steel shapes are made by passing hot steel strips (long pieces called billets, blooms or slabs) through a succession of rollers that gradually form it into the required shape, Fig. 9-8. Structural steel shapes are available in a number of sizes and weights. Fig. 9-9 illustrates most standard shapes and gives their identifying symbol and designation.

Fig. 9-7. Wood lamination has added strength and beauty to the wood construction field. (Weyerhaeuser)

Fig. 9-8. Structural steel beam is being formed by a series of rollers. (United States Steel)

A typical designation for a wide flange beam would be W12 x 50, which indicates a beam 12 inches in depth that weights 50 pounds per running foot. A typical designation for a lightweight beam, sometimes called an "I" beam, would be S24 x 120. This indicates that the beam is 24 inches in depth and weights 120 pounds per running foot.

STEEL ANGLES are used as bracings in steel framing and to construct open web steel joists, Fig. 9-10. Angles are designated with the symbol "L" and are measured along the back of the angle and thickness of the legs as: 3 x 3 x 1/2 inch and 4 x 6 x 5/8 inch.

STEEL REINFORCING BARS are also used in structural steel open web beams and in reinforced concrete to strengthen the concrete in stresses of tension (pulling or bending in floors and beams) and compression (columns). The bars are round with a projection rolled into their surface to add bond strength when the concrete sets. Bars are sized by numbers 3 to 8, representing eighths of an inch. That is, a No. 4 bar would be 4/8 or 1/2 inch in diameter. Larger bars (9 through 18) are numbered differently, beginning with No. 9 which is 1.128 inches in diameter.

WELDED WIRE FABRIC is a prefabricated material used for the reinforcement of concrete

slabs, floors and pipe. It is available in sheets and rolls.

There are two types of welded wire fabric: SMOOTH (or plain), designated by a "W"; DE-FORMED, designated "D", which has "kinks" or deformations along the wire to develop additional anchorage as the concrete sets. Previously, the fabric was specified by gauge number, and some drawings may still use this system. It is recommended that the "W" for smooth and "D" for deformed wire fabric now be used.

Welded wire fabric is further designated by numbers. An example is 6 x 8 – W8.0 x W4.0. The first number (6) gives the spacing of the longitudinal wire in inches, Fig. 9-11. The second number (8) gives the spacing of the transverse wires in inches. The first letter-number-number combination (W8.0) gives the type and size of the longitudinal wire. The second combination (W4.0) gives information on the transverse wire.

In the example given, the longitudinal wires are 6 inches apart. The transverse wires are 8 inches apart. The longitudinal wire is smooth and has a cross-sectional area of 0.08 sq. in. The transverse wire is also smooth with an area of 0.04 sq. in. Fig. 9-12 lists some of the common stock styles of welded wire fabric.

GAUGE METALS: Wall studs, lintels over doors and windows in masonry structures, window and door frames and some floor joists are made from the heavier gauge metals. Thin gauge metals are used for such items as roof flashings, duct work, roofing and wall siding.

NONFERROUS METALS: Aluminum is used to a limited extent in structural work and framing. Actually, it is an aluminum alloy since aluminum itself is very soft. It is used in its natural color for windows, doors, thresholds and siding. Aluminum may also be anodized or given a chemical finish to add color and to improve its appearance or make it more resistant to abrasion.

Aluminum alloys are made in structural shapes for use as H beams, I beams and as angles. Aluminum is also used for some duct work, screens

DESCRIPTIVE NAME	SHAPE	IDENTIFYING SYMBOL	TYPICAL DESIGNATION HEIGHT Wt/Ft in Lb.	NOMINAL SIZE HEIGHT WIDTH
WIDE FLANGE SHAPES		W	W21 X 142	21 X 13
MISCELLANEOUS SHAPES		M	M8 X 6.5	8 X 2¼
AMERICAN STANDARD BEAMS		S	S8 X 23	8 X 4
AMERICAN STANDARD CHANNELS		C	C6 X 13	6 X 2
MISCELLANEOUS CHANNELS		MC	MC8 X 20	8 X 3
ANGLES—EQUAL LEGS		L	L6 X 6 X ½*	6 X 6
ANGLES—UNEQUAL LEGS		L	L8 X 6 X ½*	8 X 6
BULB ANGLES		BL	BL6 X 3½ X 17.4	3½ X 6
STRUCTURAL TEES (CUT FROM WIDE FLANGE)		WT	WT12 X 60	12
STRUCTURAL TEES (CUT FROM MISCELLANEOUS SHAPES)		MT	MT5 X 4.5	5
STRUCTURAL TEES (CUT FROM AM. STD. BEAMS)		ST	ST9 X 35	9
TEES		T	T5 X 11.5	3 X 5
WALL TEE		AT	AT8 X 29.2	4⅞ X 7¾
ELEVATOR TEES		ET	ET4 X 24.5	4⅛ X 5½
ZEES		Z	Z4 X 15.9	6 X 3½

*SIZE ONLY

Fig. 9-9. Some examples of structural steel shapes used in construction.

Fig. 9-10. Open web steel joists being installed on a construction job. (Armco Steel Corporation)

and for electrical wiring. Copper and copper alloys are used for various construction items such as piping, flashings, roofing, screens, gutters and electrical wiring. The particular metal would be indicated on the drawing as a note and most likely detailed in the specifications in the section where it is used.

Roofing

The ROOF of a structure consists of a deck, usually wood frame and sheathing, metal or concrete. The material used to weatherproof the roof may be any number of materials, such as asphalt shingles, roll roofing, wood shingles, built-up roof of roll felt, hot asphalt or tar and gravel, or clay tile.

Roof material is designated by the pounds per square, which is the amount of material required to cover 100 square feet. An example of a designation for asphalt shingles is: 12 square of 240 pound (with color indication). Roofing materials vary in weight per square from approximately 100 pounds for roll roofing to 2900 pounds for Spanish clay tile laid in mortar.

Roofing materials usually are noted on the drawing and detailed in the specifications.

Glass

GLASS is classified as a ceramic material that melts with heat. It can be formed at temperatures above 2300 deg. F. It is made from sand (silica),

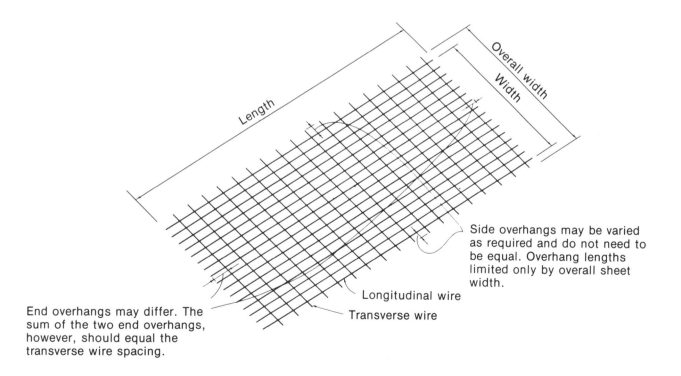

End overhangs may differ. The sum of the two end overhangs, however, should equal the transverse wire spacing.

Side overhangs may be varied as required and do not need to be equal. Overhang lengths limited only by overall sheet width.

Longitudinal wire

Transverse wire

Fig. 9-11. Welded wire fabric nomenclature. (Wire Reinforcement Institute)

Common Stock Styles of Welded Wire Fabric

STYLE DESIGNATION		STEEL AREA SQ. IN. PER FT.		WEIGHT APPROX.
NEW DESIGNATION (BY W-NUMBER)	OLD DESIGNATION (BY STEEL WIRE GAUGE)	LONGIT.	TRANS.	LBS. PER 100 S.F.
ROLLS				
6x6—W1.4xW1.4	6x6—10x10	.028	.028	21
6x6—W2.0xW2.0	6x6—8x8*	.040	.040	29
6x6—W2.9xW2.9	6x6—6x6	.058	.058	42
6x6—W4.0xW4.0	6x6—4x4	.080	.080	58
4x4—W1.4xW1.4	4x4—10x10	.042	.042	31
4x4—W2.0xW2.0	4x4—8x8*	.060	.060	43
4x4—W2.9xW2.9	4x4—6x6	.087	.087	62
4x4—W4.0xW4.0	4x4—4x4	.120	.120	85
SHEETS				
6x6—W2.9xW2.9	6x6—6x6	.058	.058	42
6x6—W4.0xW4.0	6x6—4x4	.080	.080	58
6x6—W5.5xW5.5	6x6—2x2**	.110	.110	80
4x4—W4.0xW4.0	4x4—4x4	.120	.120	85

*Exact W-number size for 8 gauge is W2.1.
**Exact W-number size for 2 gauge is W5.4.

Fig. 9-12. Common Stock Styles of Welded Wire Fabric   (Wire Reinforcement Institute)

soda (sodium oxide) and lime (calcium oxide). Other chemicals may be added to change its characteristics.

SHEET GLASS is the glass commonly used for window glass in thicknesses of 3/32 inch (single strength—SS) and 1/8 inch (double strength—DS). Thick glass (sometimes improperly called semiplate) is a sheet glass 3/16 to 7/16 inch in thickness.

PLATE GLASS is a sheet glass that has been specially treated by heat on the surface as it is formed, producing a brilliant surface which is ground and polished when it has cooled.

SAFETY GLASS has been developed to overcome the hazards of sheet glass in large, exposed or public areas. Three types of safety glass are available: tempered, laminated and wired glass.

TEMPERED GLASS is developed by heating annealed glass almost to the melting point and chilling it rapidly, creating high compression on the exterior surfaces and high tension internally. This makes the piece of glass three to five times as strong as annealed glass. Tempered glass may be broken, but it usually shatters into small, pebble-like pieces rather than sharp slivers. Tempered glass must be ordered to the exact size before tempering since it cannot be cut, drilled or ground after it has been tempered.

LAMINATED GLASS consists of layers of poly-vinyl butyral between two or more sheets of glass. The layers of plastic and glass are bonded together with heat and pressure to form a single unit. The elasticity of the plastic serves as a cushion for any object striking the laminated glass. This glass can be broken, but the plastic layers holds the small, sharp pieces in place thus avoiding the hazard of broken glass.

WIRED GLASS is a glass that has a wire "mesh" molded into its center. Wire glass may be broken, but the wire tends to hold the small parts together rather than flying or falling (as in a skylight) to injure someone. Wired glass may be obtained in an etched or sandblasted finish, or pattern effect on one side for privacy.

INSULATING GLASS is a unit of two or more sheets of glass separated by air space, which is dehydrated at atmospheric pressure and sealed. These units serve as a good insulator for heat and sound transfer. A typical insulating glass installation in a modern window sash is shown in Fig. 9-13.

Fig. 9-13. A type of window utilizing insulating glass. (Andersen Corp.)

PATTERNED GLASS is sheet glass that has had a pattern rolled into one or both sides to diffuse the light and provide privacy.

STAINED GLASS, sometimes called art or cathedral glass, is produced by adding metallic oxides when it is in a molten state. This glass may be used in sheets of cut into smaller pieces and made into leaded glass for windows or decor pieces.

SYMBOLS for glass consist of a single line on plan drawings or section small scale drawings, or the symbol may consist of several lines on large scale drawings, Fig. 9-1. Glass areas in elevation views are left plain or consist of a series of random diagonal lines indicating a sheet of glass.

Ceramic Tile

CERAMIC TILES are used for floor and wall coverings of residences and commercial buildings. The tile comes in sizes of 3/8 inch square to units of 16 x 18 inches. Popular wall sizes are 4 1/4 x 4 1/4 inches, 4 1/4 x 6 inches and 6 x 6 inches. Hexagonal and octagonal tiles are also available.

The tiles may be glazed or unglazed. Glazed tiles usually are 5/16 inch thick. Unglazed (faience) tiles vary from 7/16 inch to 3/4 inch thick.

Mosaic tiles are tiles that may be laid to form a design or pattern. They usually are 6 inches or less in size and may be glazed or unglazed.

Quarry tile are used primarily for floor coverings and are produced from clays that provide a wear resistant surface.

Tiles are set in a portland cement, latex adhesive or an epoxy mortar, then the joints between the tiles are grouted.

Symbols for tile in elevation views show a small section indicating tile, Fig. 9-1. Some drawings will show the exact pattern the tile is to be laid. Tile in section views is represented by two sets of 45 degree section lines crossing. See Fig. 9-1.

Plastics

PLASTICS have many uses in construction. Among the more common uses are:
Plastic laminates as counter tops, doors and wall surfacing.
Panels of wood or gypsum which have been printed and textured and given a plastic vinyl coating (increasingly used as wall surfaces in residential and commercial building).
Rain gutters and downspouts.
Plastic pipe for water-transmission systems, sprinkling systems and draining and sewage systems.

Plastics are also used for many trim and ornamental decor items, such as moldings on doors and panels and emblems that simulate wood carvings. Plastic usually is noted on the drawing and detailed in the specifications.

Insulation

The purpose of THERMAL INSULATION is to reduce the heat transmission through walls, ceiling and floors of buildings. Most building materials have some insulating value. The resistance to heat transmission can be increased through the use of the right insulation materials and their proper installation.

Heat is transferred in three ways: conduction, convection and radiation.

CONDUCTION is the transfer of heat when a given material is held in direct contact with another material. A ceiling which has no insulation other than the gypsum board plaster will permit far more heat loss than one insulated with the proper material.

CONVECTION is the transfer of heat by another agent, usually air. When air in a room is heated by a poorly insulated wall, it tends to rise, thus causing a circulation of air past the wall and a gain in room temperature. The reverse is true in the winter. The wall is chilled and the room cooled, requiring more heat to remain at a comfortable temperature.

RADIATION is the transfer of heat by wave motion in a manner similar to the transmission of light. Heat from the sun is radiant heat. The sun's waves heat objects which they contact, but not the space through which they move. The sun's rays will heat a dark roof surface more than a light colored or shiny, reflective surface. Reflective insulation, such as aluminum foil used in connection with other types of insulation and gypsum board, helps to reduce gain from radiation.

Heat loss or gain through walls and ceilings are a result of all three forms of heat transfer. Proper insulation materials and installation will effectively reduce the loss.

Types of Insulation Materials

Insulation is manufactured in a variety of forms and types to meet specific construction requirements. Each type will have an "R" (resistance to heat transfer) value, depending on the manner of application and amount of material. A high "R" value means good insulation qualities. Insulation materials are classified as:
1. Flexible (blanket or batt).
2. Loose-fill.
3. Reflective.
4. Rigid (structural and nonstructural).

FLEXIBLE insulation is available in blanket and batt form, Fig. 9-14. Blankets come in widths suitable to fit 16 and 24 inch stud and joist spacing and thicknesses of 1 to 3 1/2 inches. The body of the blanket is made of mineral or vegetable fiber

such as rock wool or glass wool, wood fiber and cotton. Organic materials are treated to resist fire, decay, insects and vermin. Blanket are covered with a paper sheet on one side and a vapor barrier of aluminum foil or asphalt on the other. The vapor barrier is installed facing the warm side of the wall.

Batt insulation is made of the same material as blankets, in thicknesses of 2 to 6 inches and lengths of 24 and 48 inches.

LOOSE-FILL insulation is available in bags or bales. It is either poured, blown or packed in place by hand. Loose-fill insulation is made from rock wool or glass wool, wood fibers, shredded redwood bark, cork, wood pulp products, vermiculite, saw dust and shavings. This insulation is suited for insulating sidewalls and attics of building. It is also used to fill "cells" in block walls during construction.

REFLECTIVE insulation is designed to reflect radiant heat. It is made from aluminum foil, sheet metal with tin coating and paper products coated with a reflective oxide composition. To be effective, the reflective surface must face an air space of at least 3/4 inch. When the reflective surface contacts another material, such as a wall or ceiling, the reflective properties are lost along with insulating value. This material is often used on the back

Fig. 9-14. Flexible insulation being installed above ceiling. Vapor barrier should always be installed facing warm-in-winter portion of home.   (Owens-Corning Fiberglas)

of gypsum lath and blanket insulation.

RIGID insulation usually is made of a fiberboard material in sheet form. Common types are made from processed wood, sugarcane and other fiber. These produce a lightweight, low-density product with good heat and acoustical insulating qualities.

Rigid insulation is used as sheathing for walls (structural) and roof decks (nonstructural) as an added insulating factor with other forms of insula-tion materials. Fig. 9-1 shows symbols for insulation.

A recently developed product is the energy-saving AIR INFILTRATION BARRIER. It is a sheet of polyethylene fibers placed over sheathing. This keeps air out of the wall cavity by sealing cracks and seams. Heat loss through walls is reduced by preserving the insulation's R-value.

An infiltration barrier is noted on the plan (see Fig. 8-10) and detailed in the building specifications.

Blueprint Reading Activity 9—1
CONSTRUCTION MATERIALS

After studying the materials in Unit 9, answer the following questions.

1. What causes concrete to "set" or harden? How long does this take?

1. _____

2. Why are admixtures sometimes added to the concrete mix?

2. _____

3. What characteristic of concrete is strengthened by placing reinforcing steel in the concrete?

3. _____

4. Explain the principle of prestressed concrete which greatly increases its strength.

4. _____

5. Stones are used chiefly for what purpose in construction today?

5. _____

6. Bricks are a manufactured product in contrast to stones. What type bricks are most often used in construction?

6. _____

7. How does mortar for stone masonry differ from mortar for brick masonry? Why?

7. _____

8. What materials are used as back up in composite masonry walls?

8. _____

9. Gypsum blocks are used chiefly for what purposes?

9. _____

10. Distinguish between the following wood products:
    a. Lumber.
    b. Rough framing lumber.
    c. Finish lumber.
    d. Plywood.

10. a. _____

    b. _____

    c. _____

    d. _____

11. How is a framing member in section indicated on a drawing?

11. _____

12. What is a glue-laminated structural wood member?

12. _____

13. Define the term structural steel.

13. _____

14. How are steel reinforcing bars sized and designated?

14. _____

15. Interpret the following welded wire fabric (WWF) specification: 6 x 6 - W2.9 x W2.9.

15. _____

16. How are roofing materials designated?

16. _____

17. Explain the difference between sheet glass and plate glass.

17. _____

18. There are three types of safety glass. Describe these.

18. _____

19. Name some common uses of plastics in construction.

19. _____

20. Heat is transferred in three ways. Explain these and give an example of each in construction.

20. _____

Unit 10
Specifications In Construction

SPECIFICATIONS are written statements that define the quality of work to be done and the materials to be used in construction. They supplement the drawings which, in turn, describe the physical location, size and shape of the construction project.

In this unit, the makeup of specifications in construction is explained. Also, a typical set of specifications is presented to acquaint you with its organization and to assist you in locating information concerning a particular phase of construction.

Requirements and Scope

Specifications are of primary interest to the architect and the owner as well as the general contractor and subcontractors who will bid on the construction project. Specifications also are important to the various craftsworkers engaged in the construction, since they are directly performing the work which must conform to "specs." The following statements are excerpts from a typical set of specifications:

General Requirements:

Work shall include all items (building and site) indicated on these drawings unless otherwise noted. The American Institute of Architects (AIA) General Conditions, latest edition, are hereby a part of this contract . . . General contractor shall remove all construction debris from the job site and leave the building broom clean.

Structural Specifications:

All footings shall be carried to undisturbed medium sand or rock, and to at least the depth shown on the drawings.

All concrete to attain a minimum ultimate compressive strength of 3000 PSI in 28 days. Aggregates to be clean and well graded maximum size of 1". Concrete slump 3" minimum to 5" maximum.

Specifications, along with the drawings, assure that the construction project will be completed in the manner it was intended.

The Elements

Specifications differ to some extent for various construction projects. However, there are certain ELEMENTS which are rather common to any construction job. The elements defined by The Construction Specifications Institute are:
- Contract Requirements (Documents)
- General Requirements
- Sitework
- Concrete
- Masonry
- Metals
- Wood and Plastics
- Thermal and Moisture Protection
- Doors and Windows
- Finishes
- Specialties
- Equipment

- Furnishings
- Special Construction
- Conveying Systems
- Mechanical
- Electrical

Coverage

Specifications vary in length from a few pages to hundreds of pages for more complicated projects. Specifications normally are prepared by a person or persons in the architect's office who are thoroughly familiar with construction procedures, building materials and local government codes.

Specifications and drawings supplement each other, so that all information needed for a construction project is included. An example would be the hardware for use on doors. The drawings would show location and direction of door swing; the specification would indicate quality or brand name, style, finish and any other information to assure the intended type of hardware is included.

An example of a specification covering standards for quality of work to be done in the bricklaying craft would be: "Brick shall be laid true to lines and plumb; mortar joints shall be completed filled."

Purposes Served

Specifications serve to clarify information that cannot be shown on the drawing. To put all the information needed on construction drawings to "spell-out" work standards, type of materials to be used, and the responsibility of various parties to the contract would make the drawings too confusing and nearly impossible to use. Specifications and drawings each provide information on details of construction. When a difference occurs between the specifications and the drawing, the specification holds. This is usually stated in the general conditions section of the specifications. In any event, such a difference should be called to the attention of your supervisor.

Specifications serve these major purposes:

1. As a guide for general contractors as well as subcontractors in bidding on a construction job. Without such specifications it would be difficult to evaluate the bids of various contractors.
2. As a standard for quality of material and work to be done in a construction project and to assure the owner of getting what he ordered.
3. As a guide for the building engineering and inspection departments of the local government in checking for compliance with building codes and zoning ordinances.
4. As the basis of agreement between the owner, architect and contractors in settling any disputes that may arise regarding the construction project.

Types of Information Provided

Specifications, and the information contained in them, vary from a simple outline (for "stock plan" houses) that contains minimal information on material and work to be done to elaborate sets of specifications that detail every item of material and work to be performed.

One type of information contained in specifications is referred to as contract requirements or "documents." These cover the NON-TECHNICAL ASPECTS such as terms of contract agreement, responsibilities for examining the construction site, insurance, permits, utilities, and supervision of construction. It is in these areas of specifications that misunderstandings can easily occur. Therefore, these requirements usually are more detailed on complex construction jobs.

A second type of information contained in specifications relates to the TECHNICAL ASPECTS of the construction project. This information is listed by major divisions, usually in the order the work is performed on the job. Typically, these divisions are: Sitework, Concrete, Masonry, Metals, Wood and Plastics, and on through the various aspects of the construction project. Materials are specified by standard numbers, such as ASTM (American Society of Testing and Materials) or grades of lumber specified by lumber manufacturer's associations.

In addition to specifying materials, the technical section of the specifications also states the standards for quality of work to be done, or how materials are to be installed or applied.

How to Read Specifications

Specifications usually are written around some standard format proposed by the American Institute of Architects or the Construction Specifications Institute. When reading specifications, first review the Table of Contents to become familiar with the type of information included and to get an overview of the organization of the specifications.

Information of a general nature, such as insurance coverage, supervision of construction, and responsibilities of various parties would be found under the "General Requirements" section of the specifications.

To locate the specification for a certain item, say the "concrete mix to be used in the foundation walls," locate the item under its classification in the Table of Contents or Index. For concrete mix, the obvious classification would be "Concrete." Next, check the subheading dealing with the type of concrete ("Cast-In-Place Concrete"). Information on "framing of exterior walls" most likely would be found under "Wood and Plastics."

A portion of a complete set of specifications for a light commercial building is included at the end of this unit to give you some experience in reading specifications. The following assignment will assist you in becoming familiar with the organization of a set of specifications and in locating information concerning a particular phase of the construction.

Blueprint Reading Activity 10−1
READING CONSTRUCTION SPECIFICATIONS

Refer to the partial set of specifications at the end of this assignment, and to the CSI Specifications beginning on page 243, to answer the following questions.

1. Into how many divisions, other than "Contract Requirements (Documents)," is this set of specifications divided?

1. _____

2. Under what division would you expect to locate:
   a. Roofing materials and processes?
   b. Finish trim?
   c. Interior finishing of walls?

2. a. _____
   b. _____
   c. _____

3. What must all bidders do before submitting a bid?

3. _____

4. How do the specifications relate to the drawings on this construction job?

4. _____

5. If an item is mentioned in the specification and not in the drawings, must the bidder supply it?

5. _____

6. Who has the final decision as to the interpretation of the drawings and specifications?

6. _____

7. To whom does the term contractor refer?

7. _____

8. Who is responsible for providing and paying for temporary electrical service?

8. _____

9. Once a contractor has approval on shop drawings from the architect, is he permitted to make minor deviations from the drawings as he sees necessary without further approval?

9. _____

10. Who shall provide the necessary temporary heat needed for materials, water, etc., or any other heat required to accomplish the work?

10. _____

11. If the permanent heating plant is used for temporary heat, who will pay for its operation?

11. _____

12. Who may provide and maintain on the premises temporary storage sheds for storage of all materials which may be damaged by the weather?

12. _____

13. What is the specification controlling the flat roofing?

13. _____
_____
_____

14. How much of an overlap must be made between the new and existing roof?

14. _____

SPECIFICATIONS FOR AN OFFICE ADDITION
For GOODHEART-WILLCOX COMPANY, INC.
At 123 Taft Drive, South Holland, Illinois

PLAN NO. 12275
Architects: Donald T. Smith & Associates
7227 W. 127th Street, Palos Heights, Illinois

INDEX

SUPPLEMENTARY GENERAL CONDITIONS

General:

These Supplementary General Conditions and the Specifications bound herewith shall be subject to all the requirements of the "General Conditions of the Contract for the Construction of Building", latest edition, Standard Form of the A.I.A., except that these Supplementary General Conditions shall take precedence over and modify any pages or statements of the General Conditions of the Contract and shall be used in conjunction with them as a part of the Contract Documents. The General Conditions of the Contract are hereby, except as same may be inconsistent herewith, made a part of this Specification, to the same extent as if herein written in full.

Copies of the General Conditions of the Contract are on file and may be referred to at the Office of the Architects.

Scope of Work:

The work involved and outlined by these Specifications is for the construction work for the completion of the Office Addition for Goodheart-Willcox Co., Inc., 123 Taft Drive, South Holland, Illinois, as further illustrated, indicated, or shown on the accompanying drawings, dated 8 December 19——.

Examination of Site:

Before submitting a proposal for this work, each bidder will be held to have examined the site, satisfied himself fully as to existing conditions under which he will be obligated to operate in performing his part of the work, or which will in any manner affect the work under this contract. He shall include in his proposal any and all sums required to execute his work under existing conditions. No allowance for additional compensation will be subsequently in this connection, in behalf of any contractor, or for any error or negligence on his part.

Drawings and Specifications:

These Specifications are intended to supplement the Drawings, the two being considered complementary, the therefore, it will not be the province of these Specifications to mention any portion of the construction which the drawings are competent to explain and such omission will not relieve the contractor from carrying out such portions as are only indicated on the drawings. Should the items be required by these Specifications which are not indicated on the drawings, they are to be supplied, even if of such nature that they could have been indicated thereon.

Any items which may not be indicated on the drawings or mentioned herein, but are necessary to complete the entire work, as shown and intended, shall be implied and must be furnished in place.

The decision of the architects as to the proper interpretation of the Drawings and these Specifications shall be final and shall require compliance by the contractor in executing the work.

Figured dimensions shall have precedence over scale measurements, and details over smaller scale general drawings. Should any

Principles and Definitions:

Where the words "approved", "satisfactory", "equal", "proper", "ordered", "as directed", etc., are used, approval, etc., by architects is understood.

It's understood that when the word "Contractor" is used in these Supplementary General Conditions, the work described in the paragraph may apply to all Contractors and Sub-contractors involved with the work.

Temporary Facilities:

Temporary Heat: Each contractor shall provide the necessary temporary heat needed for materials, water, etc., or any other heat required to accomplish his work. If the permanent heating plant is used for temporary heat, the owner will pay for its operation.

Temporary Light & Power: The general contractor shall arrange and pay for a temporary electrical service taken from the existing building for use by all trades, include 60 amp. 2 pole, 4 outlet fused panel mounted on pole or wall at the site. The electrical contractor shall provide temporary light within the structure as necessary and directed. General contractor will pay for all temporary electrical service until such time when permanent meter shall be installed.

Temporary Sheds for Storage: The general contractor may provide and maintain on the premises, where directed, watertight storage shed, or sheds, for storage of all material which may be damaged by the weather. These sheds shall have wood floors raised above the grounds.

Division 7

THERMAL AND MOISTURE PROTECTION

General Conditions:

This contractor shall read the General Conditions and Supplementary General Conditions which are a part of these Specifications.

Scope of Work:

Furnish all materials and labor necessary to complete the entire roofing as shown on the drawings or hereinafter specified.

Roofing:

Flat roofing shall be 4 ply tar and gravel Spec. 102 of the Chicago Roofing Contractors Assn. except single pour on gravel.

Carry new roofing onto existing roof a minimum of 3'-0" and roof new saddles on existing.

Rigid Insulation:

Cover all flat roof surfaces over steel deck with 2" of rigid insulation.

Insulation board shall be Fesco or equal.

Insulation shall be installed according to Spec. 102 of the Chicago Roofing Contractors Assn. Form saddles on existing roof of rigid insulation so pitch is to new roof drain.

Unit 11
Reading Plot Plans

A building can be greatly enhanced by the manner in which it is located on a plot. Therefore, architects usually try to take advantage of the slope of the land, surrounding trees, view from the street, and other features to increase the appearance value of the structure.

Plot Plans

A PLOT PLAN (sometimes called a site plan) is a view from above the property which shows the location of the building on the lot. See Fig. 11-1. Other features may be shown, such as:
1. Lot and block number or address
2. Bearing (direction) and length of property lines
3. North direction point
4. Dimensions of front, rear and side yards
5. Location of other accessory buildings (carport, garage, etc.)
6. Location of walks, drives, fences and patios
7. Location of easement setbacks
8. Location of utilities
9. Elevation levels at:
   a. First floor of building
   b. Floor of accessory buildings
   c. Finish grade at each principal corner of the building
   d. Existing and finish grade at each corner of the plot
10. Trees and shrubs to be retained
11. Scale of the plot plan

North Point Arrow

Most architectural plot plans will have an ARROW INDICATING TRUE NORTH for purposes of further identifying the lay of the plot and orientation of the building. See Fig. 11-1. The design of a building is frequently planned to minimize or take advantage of the sun's rays. Some architects use a second directional arrow when the building does not align closely with the major directions. The purpose of this arrow is to simplify the reference to the various elevations such as NORTH ELEVATION or EAST ELEVATION.

Property Lines

Lines outlining the building plot are called PROPERTY LINES, and the length and bearing (direction) of each line is identified on the plot plan, Fig. 11-1. The first corner or point-of-beginning (P.O.B.) is identified as a certain distance and bearing (direction) from an easily recognizable monument (permanent feature), such as a survey marker or a manhole. NOTE: Monuments are marked on county and city maps showing roads or streets. The plot plan includes surveyor's dimensions of each property line in feet and decimal hundredths of a foot. Bearings are given in degrees, minutes and seconds.

A bearing is read clockwise from the NORTH or SOUTH POINT. In Fig. 11-1, the first property line

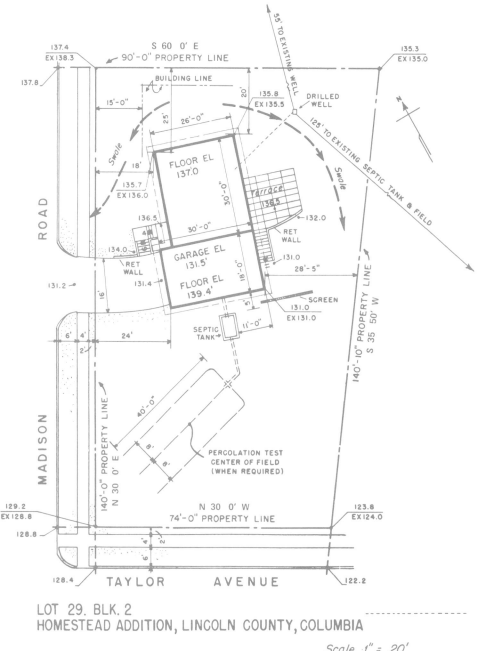

Fig. 11-1. A plot plan locates the building on the property.

is N30°00'E 140'-0'' feet. This would be 140 feet from the P.O.B. in a direction 30 degrees from true north toward the east. The bearing of a line is taken from the north or south azimuth, whichever keeps the angle 90 degrees or less.

When the property line is a curve rather than a straight line, it is identified by a radius, length of curve and its angle of tangency: for a DELTA (Δ) reading in degrees, minutes and seconds. A delta is the central angle formed by the radii meeting the

curve at the point of tangency and used in calculating the length of the curve.

Contour Lines

Drawing with lines that show elevations (heights) of the plot are called CONTOUR MAPS OR TOPOGRAPHIC DRAWINGS, Fig. 11-2. The elevations on the particular plot are referenced to a local permanent monument easily accessible, such as a survey marker plate, a fire hydrant or manhole

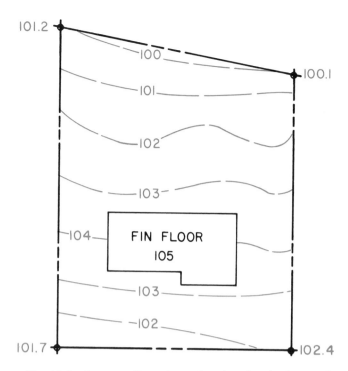

Fig. 11-2. Contour lines show the elevation in feet and the general lay of the plot.

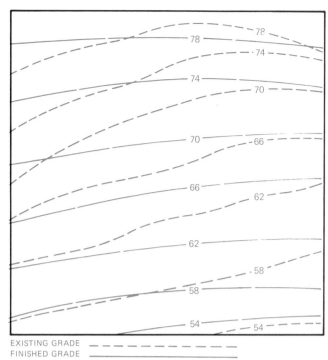

EXISTING GRADE
FINISHED GRADE

Fig. 11-3. Contour lines showing the original and finished grade for a plot.

cover. These elevations readings may be actual readings of distances above sea level. To simplify the reading and laying out of elevations, some architects assign a "zero" point such as 100'-0" to an elevation on the plot or building and reference other elevations from this.

Contour lines are drawn on the plot plan to indicate the topography (changing level of land) of the plot and to assist in selecting the best location for a building. The interval (vertical distance) of the contour lines may represent any convenient distance such as 2, 5 or more feet depending on the size and slope of the plot. Lines which are far apart indicate a gradual slope of the land; lines close together indicate a steep slope. The elevation of each line is indicated in feet above or below the "zero" point and all points on the line are at that elevation.

Contour lines are long, freehand dash lines, Fig. 11-3. When it is desired to show both the original grade and a finish grade of contour, the original is shown in short dash lines.

Location of Building on Plot

The exterior lines of the building are located by measuring in from the property lines as indicated on the blueprint. Stakes are driven in the ground at the corners of the building and a small nail driven in the top of each stake to locate the exact corner. See Fig. 11-4.

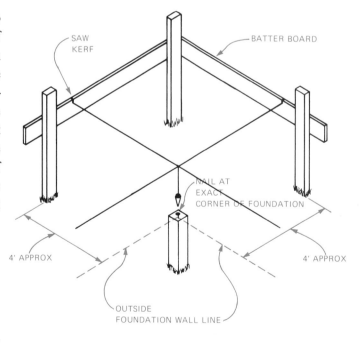

Fig. 11-4. Stake with nail to locate exact corner of foundation wall.

Batter boards are erected approximately four feet back from each corner to permit excavation of the footings and foundation, Fig. 11-5. The batter boards should all be erected at approximately the same level.

Next, using a plumb bob, lines are stretched over the batter boards so they line exactly with the corners of the building. Then, saw kerfs are cut in the batter boards to mark the exact location of the lines. Refer again to Fig. 11-4.

Topographic Features

Some blueprints show the TOPOGRAPHICAL (description of natural and manufactured) features of a plot, including what features are to be retained, changed or added. Fig. 11-6 illustrates the more common topographical symbols used on plans. Other features which appear on blueprints should be identified in a legend on the print.

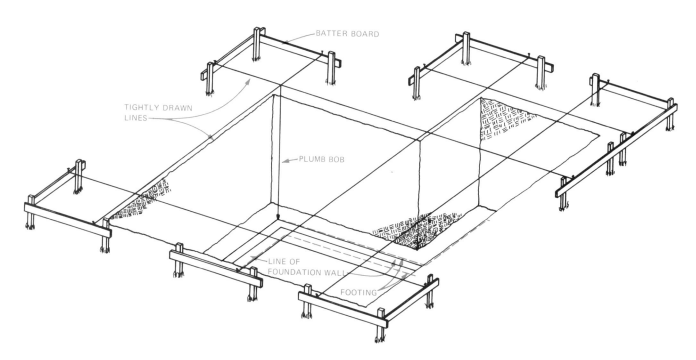

Fig. 11-5. Locating footing and foundation wall by use of lines and plumb bob.

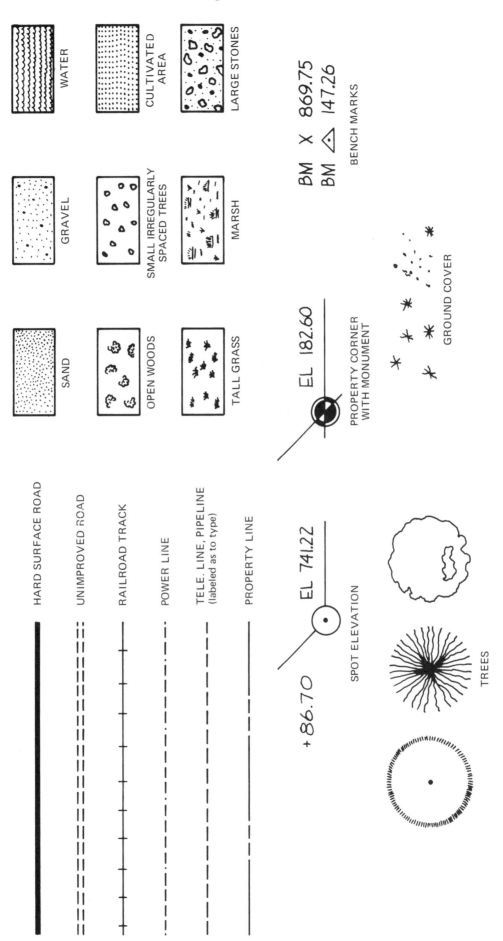

Fig. 11-6. Common topographical symbols used on plot plans.

Refer to Print 11-1 on page 103 to answer the following questions.

1. What is the scale of the plot plan?

1. _____

2. Starting with the property corner marked A, give the bearing and length of each property line.

2. A to B _____

B to C _____

C to D _____

D to A _____

3. The location of the residence is not aligned with True North. The front of the residence faces what assigned direction?

3. _____

4. What distance is the house set back from the street?

4. _____

5. What features are shown on the plot plan other than the residence itself?

5. _____
_____
_____
_____
_____
_____
_____
_____

6. Make a simple sketch, showing building centered on lot and properly set back from street.

6.

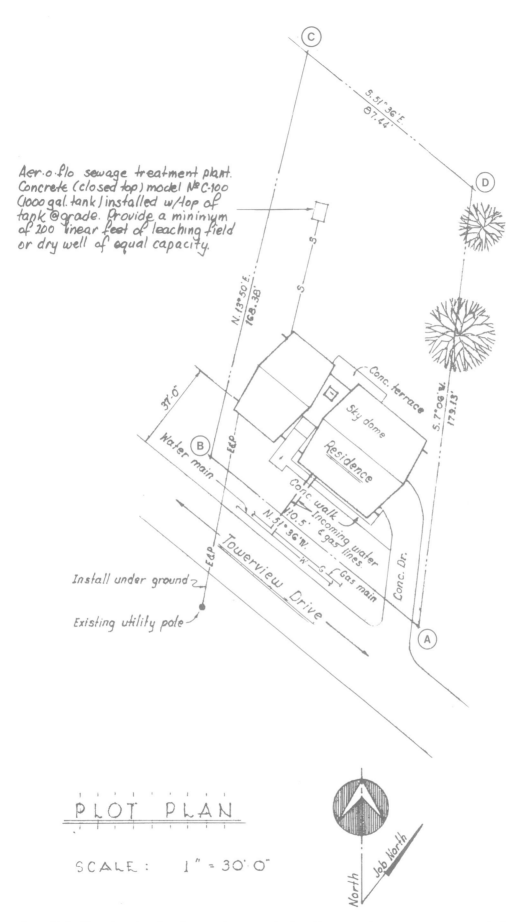

Aer·o·flo sewage treatment plant.
Concrete (closed top) model № C-100
(1000 gal. tank) installed w/top of
tank @ grade. Provide a minimum
of 200 linear feet of leaching field
or dry well of equal capacity.

S. 51° 36' E.
87.44'

N. 13° 50' E.
168.38'

37'·0"

S. 7° 06' W.
179.13'

Conc. terrace

Sky dome

Residence

Water main

Conc. walk

Incoming water

N. 51° 36' W.
110.5'

& gas lines

Gas main

Conc. Dr.

Towerview Drive

Install under ground

Existing utility pole

PLOT PLAN

SCALE: 1" = 30'·0"

North

Job North

Print 11-1. The plot plan for the residence studied in Unit 11.   (Batson & Associates Architects)

Refer to Print 11-2 in the Large Prints Folder to answer the following questions.

1. What is the scale of the plot plan?   1. _____

2. How many sheets are there in this set of plans?   2. _____

3. Give the reading for each property line, starting   3. A to B _____
   with A to B.   B to C _____
                                                            C to D _____

                                                        _____

                                                        D to E _____

                                                        _____

                                                        E to A _____

4. The front of the building would be called the   4. _____
   _____ elevation based on the north
   orientation arrow.

5. Explain what shaded areas F and G are and   5. _____
   what is to be done.   _____
                                                         _____
                                                         _____
                                                         _____

6. A curb bounds the new asphalt paving on the   6. _____
   north. What bounds it on the east? Where
   would the detail for this be found?

7. Where would a detail of the concrete curb J be   7. J_____
   found? Curb K?   K_____

8. What is the level of finish grade in planter H?   8. _____
                                                         _____

9. How many parking spaces are to be provided   9. _____
   along the east side and how wide is each?   _____

10. Which driveway low point at entry, south or   10. _____
    west is the higher after finish grade?

11. How wide are the driveways at the curb entry?   11. _____

12. What is the existing grade and finished floor   12. Existing_____
    level for the New Building?   Finished Floor_____

Unit 12
Concrete Footings, Foundations
and Floors Blueprints

Once the building has been located on the plot and the excavation is complete, work begins on the concrete footings and foundation walls. The details of construction for the footings and foundation walls are contained in the Foundation Plan or Basement Plan (if the building has a basement).

The footings and walls must be carefully laid out since the remainder of the entire structure depends on the accuracy of this important preliminary step of the construction procedure. To make blueprint reading easier, this unit explains how footings and foundations are located, excavated and constructed. In addition, slab floor construction is covered.

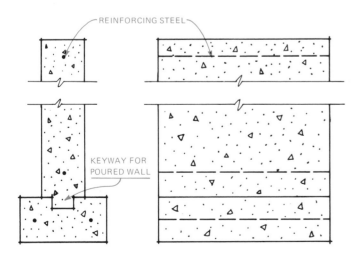

Fig. 12-1. Footings are designed to carry the weight of the entire building.

Footings

FOOTINGS are the "feet" placed in the ground upon which the foundation walls and entire building load rests. The size of footings are shown on the foundation plan or on a detail of the foundation wall. See Fig. 12-1. The footing size is determined by architects and engineers, based on the type of soil (determined by tests) and the weight of the building.

Footings also are required under support columns. These footings frequently are wider and thicker than footings for foundation walls because the column loads are concentrated in one spot. See Fig. 12-2. Fireplace chimneys and similar concentrations of weight also require larger footings.

Footings are located from strings attached to batter boards that are set back from the excavation. The footings may be trenches cut into the floor of the excavation or they may rest on the excavation floor. In the latter case, boards are used to form the sides to the proper width and adjusted for the correct height.

Footings must rest on UNDISTURBED EARTH, and steel reinforcing rods usually are placed in the footings. This is especially important where footings must pass over previously disturbed earth due to the installation of a drain pipe or other excavation. When a poured concrete foundation wall is to be erected on the footing, the drawing may call for

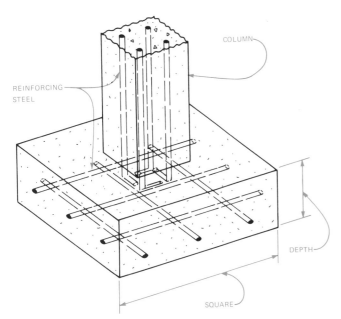

Fig. 12-2. Footings for columns have a larger cross section than wall footings.

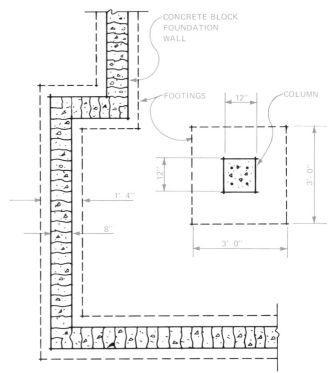

Fig. 12-3. Plan view of footings for wall and column.

a "keyway" to be cast in the footing to anchor the wall. See Fig. 12-1.

SYMBOLS for footings on a foundation plan are hidden lines, Fig. 12-3, showing the width of the footing under the foundation walls and columns. Reinforcing rods are shown as dots in sectional views. Refer again to Fig. 12-1. On elevation drawings, these rods are indicated by long dash lines. In addition, notes on the drawing and given in the specifications should be carefully checked for details relating to construction of the footings.

Foundations

FOUNDATION WALLS are the base of the building. They transfer the weight of the building to the footings and earth below. Foundation walls may be poured concrete or concrete masonry units (concrete block). Poured concrete is used where soil and weather conditions place considerable side pressure on the walls. Where applicable, concrete masonry units are an efficient way of constructing a foundation wall since no forms are required. Both types of walls usually are reinforced with steel rods.

Foundation walls and columns are shown as solid lines on the foundation plan and on the basement plan, Fig. 12-3. The space between the lines represents the material used. Foundation walls and footings are shown as hidden lines in the elevation views, Fig. 12-4. A foundation wall

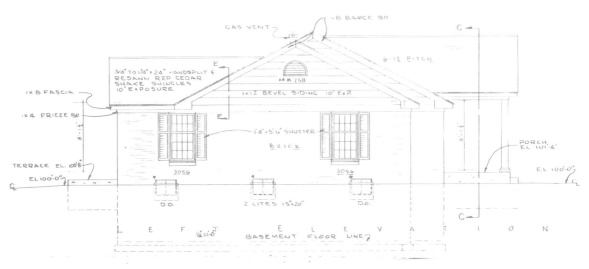

Fig. 12-4. Elevation view of a residence showing foundation wall and footing as hidden lines.

section is shown in Fig. 12-5. The material symbols used in the section indicate general type of material used. However, these materials will be detailed in notes on the drawing or in the specifications.

Fireplaces and chimneys are shown on the foundation plan with appropriate dimensions. Detail plan views and profile sections usually are drawn to provide construction details, Fig. 12-6.

Slab Floors

Concrete floors poured at ground level are called SLAB FLOORS. Concrete slabs are used as basement floors and as main floors in some areas of the

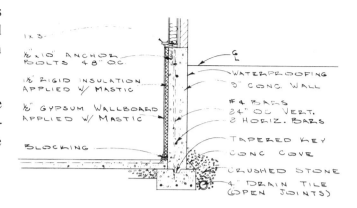

Fig. 12-5. A foundation wall section provides a detailed view of the footing and wall construction. (Garlinghouse Plan Service)

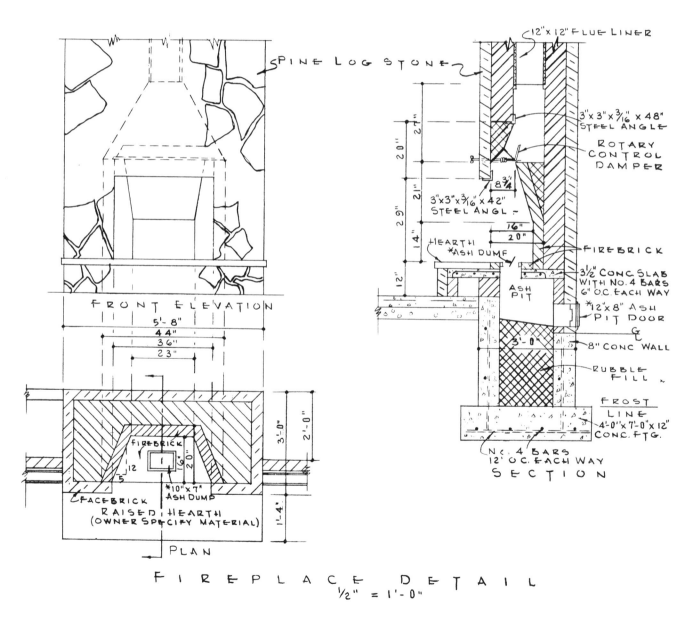

Fig. 12-6. Sectional view of fireplace provides necessary construction details and dimensions.

United States where frost penetration is no problem. Basement floors are poured after the footings and foundation walls are in and, sometimes, before the rough framing starts.

Foundations with a monolithic (one continuous unit) slab are called "floating slab" construction, Fig. 12-7. Another method of pouring a slab floor is to pour the foundations walls to floor height. Then, the floor is poured within the walls, separated by a 1/2 inch expansion joint. See Fig. 12-8. Load bearing partition walls over slab floors call for a thickened area, simulating a footing. These areas are indicated by hidden lines and a note, Fig. 12-9.

REINFORCING SLAB FLOORS: When a concrete slab is expected to be subjected to tension (due to the settling of a dirt fill or heavy load), steel reinforcing rods or welded wire fabric are cast in the concrete. A typical note specifying welded wire fabric in a concrete floor would read:

5" CONC - 6x6 - W10xW10 WWF OVER 4" ABC

Waterproofing Foundations

WATERPROOFING of foundation walls is called for in some areas where soil and climatic conditions demand protection from below grade water. Waterproofing usually consists of mopping the outside of the foundation wall with tar or asphaltum and, sometimes, the application of a polyethylene sheet over the tar. Foundations to be waterproofed will have a heavy black line on the exterior wall with a note indicating location. Also, the building specifications may specify the exact material and process.

A layer of crushed rock or gravel usually is laid below the floor area with a heavy plastic vapor barrier to keep the dampness in the ground from transferring to the slab floor.

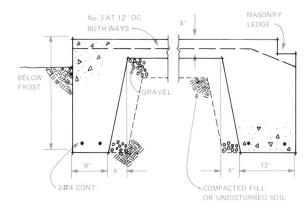

Fig. 12-7. A monolithic slab foundation.

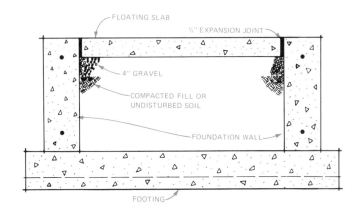

Fig. 12-8. A slab floor poured within.

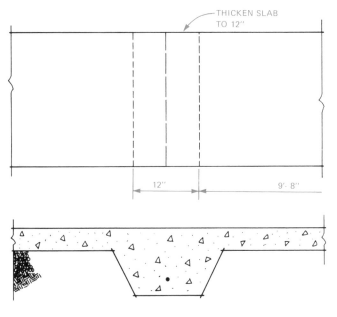

Fig. 12-9. A "footing" for an interior load bearing wall.

Concrete Footings, Foundations and Floors Blueprints

CONCRETE FOOTINGS, FOUNDATIONS AND FLOORS FOR A RESIDENCE

Refer to Print 12-1a in the Large Prints Folder and Print 12-1b on page 111 to answer the following questions.

1. What is the scale of the plan drawing?   1. _____

2. Give the overall length and width dimensions of the basement floor plan.

    2. Length _____
       Width _____

3. Give the following:
   a. Footing dimensions
   b. Number and size of reinforcing bars in footing
   c. Fireplace footing (utility room) dimensions
   d. Foundation wall thicknesses
   e. Foundation wall height above footing
   f. Number and size of reinforcing bars in wall.

    3. a. _____
       b. _____
       c. _____
       d. _____
       e. _____
       f. _____

4. How many windows are there, and what is their size.

    4. _____
       _____

5. Inside stair dimensions:
   a. Stairwell width
   b. Number of treads
   c. Length of tread run
   d. Dimension for "X"

    5. a. _____
       b. _____
       c. _____
       d. _____

6. Outside stair dimensions:
   a. Stairwell width
   b. Number of treads
   c. Length of tread run

    6. a. _____
       b. _____
       c. _____

7. Give the inside dimensions of the basement room.

    7. Length _____
       Width _____

8. Give the following dimensions of the cistern:
   a. Exterior overall length
   b. Face of exterior foundation wall to exterior cistern wall.
   c. Exterior wall thickness.
   d. Cinder block filter wall thickness

    8. a. _____
       b. _____
       c. _____
       d. _____

9. What are the inside dimensions of the bath?

    9. Length _____
       Width _____

10. What is the hall width?

    10. _____
       _____

11. What is the distance between the foundation walls along the hall and inside stairway?

11. _____

12. What is the thickness of the concrete floor slab?

12. _____

13. What separates the floor slab from the foundation wall?

13. _____

14. How is the exterior face of the foundation wall to be treated?

14. _____

15. What size drain tile is to be installed along the footing?

15. _____

HAND SPLIT CEDAR SHAKE WOOD SHINGLES
1" RIGID INSULATION
2¼" LAMINATED WOOD DECK.

12"

2"

METAL GUTTER
FLASHING
1X8 WOOD FASCIA
COVE MOLD

2X8 T&G SHEATHING
5/8" TEXTURE III
(VERT.) SIDING

4'-0"

4X10 VERT. LAMINATED WOOD BEAM
2X4 WOOD BLOCKING
COVE MOLD
2X4 WOOD PLATE
(2) 2X10 CONT. WOOD BEAM
½" DRYWALL

HEAD

11⅛"

WOOD (PROJECTING OUT) WINDOW

SILL

6'-10¼"

4" INSULATION
WOOD BASE
WOOD SHOE
CARPET (OVER ½" PLY. WD.)

5/8" PLY. WD. SUB FL.
2X10 HEADER
2X8 T&G SHEATHING
2X10 JOIST 16" O.C.
2X4 PLATE

MASONRY BEYOND
BRICK SILL
FLASHING
WEEP HOLES
GROUND LINE

10" POURED CONC.
HORZ. REINF. 3-#3 BARS
VERT. REINF. #3 @ 6'-0" O.C.
½" EXPANSION JOINT
4" CONC. SLAB @ 21" MESH

WATER-PROOF
(KEEP BELOW GRADE)

4" BANK RUN
VAPOR BARRIER
4" TILE
24" X 8" CONC. FTG
3 #3 BARS

TYPICAL WALL SECTION

SCALE 1½" = 1'-0"

Print 12-1b. Typical wall section of the Footing and Foundation blueprints.

Refer to Print 12-2a in the Large Prints Folder and Print 12-2b on page 114 to answer the following questions.

1. What is the scale of the drawing on Sheet 3?

   1. _____

2. Give the overall length and width dimensions of the building foundation.

   2. Length _____
      Width _____

3. What are the floor plan exterior dimensions of the vault?

   3. _____

4. List the vault footing dimensions and floor thickness.

   4. Footing: _____
      Floor: _____

5. What are the wall and ceiling thicknesses of the vault?

   5. _____

6. What reinforcing is specified on the drawing for the vault floor, walls and ceiling?

   6. _____
      _____
      _____

7. Give the rough opening of the door to the vault.

   7. _____
      _____

8. What are the dimensions of the typical footing beneath the exterior wall?

   8. _____

9. How far below natural grade are footings to rest and what type soil?

   9. _____
      _____

10. What reinforcing steel specifications are given for this footing?

   10. _____
       _____
       _____

11. What vertical reinforcing rods are to be installed and at what spacings?

   11. _____

12. What is called for at A?

   12. _____
       _____
       _____

13. What size footings are required for the columns in front?

   13. _____

14. What size footings are detailed for the Pylons on the North side of the building? Reinforcing steel?

   14. _____
       _____
       _____

15. What material is used for the foundation wall, and how is it to be reinforced?

15. _____
_____
_____

16. How is the foundation wall to be treated at the door and window openings?

16. _____
_____

17. Give the following for the floor:
    a. Finished floor elevation
    b. Thickness
    c. Gravel beneath
    d. Expansion joint

17. a. _____
    b. _____
    c. _____
    d. _____

18. How does the floor differ at openings for:
    a. Doors

    b. Windows

18. a. _____
    _____
    _____

    b. _____
    _____

19. What is called for at B?

19. _____
_____

20. Give the dimensions of the sidewalk thickness and edge detail.

20. Thickness: _____
    Edge: _____

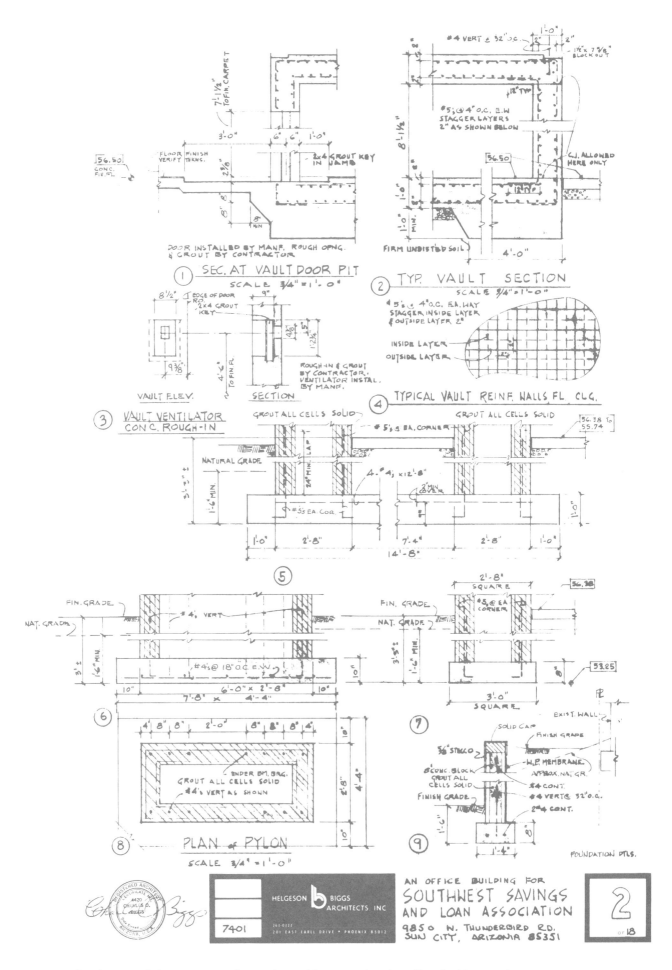

Print 12-2b. Sheet 2 of the Footing and Foundation blueprints for the office building in this assignment.

Unit 13
Reinforced Concrete and
Structural Steel Framing Blueprints

Large commercial and industrial buildings are constructed around a framework of reinforced concrete or structural steel, Fig. 13-1. In some cases, this type of construction is used in larger residences and light commercial buildings.

Concrete and steel as building materials were discussed in Unit 9; footings and foundations were presented in Unit 12. The purpose of this unit is to assist you in understanding the special applications of reinforced concrete and structural steel. Also, a careful study of this unit will assist you in reading blueprints relating to structural framing.

Reinforced Concrete

Concrete is a valuable building material if considered only for slab floors, drives and walks. Its applications become even more valuable when combined with steel reinforcement in structural applications. The compression strength of concrete and the tension strength of steel make REINFORCED CONCRETE a unique building material. Indeed, if it were not for reinforced concrete and structural steel, many of today's buildings would have been impossible to construct.

FLOOR AND ROOF SYSTEMS are of two general types, the one-way system and two-way system. The ONE-WAY FLOOR SYSTEM is so named because the floor slab is supported by girders which run parallel (one way) and rest on columns. For heavier construction carrying greater

Fig. 13-1. Structural steel and reinforced concrete are used for framework in large construction projects. (Armco Steel Corporation)

floor loads, this one-way system sometimes uses beams running at right angles to the one-way girders. See Fig. 13-2. The floor slab and girder

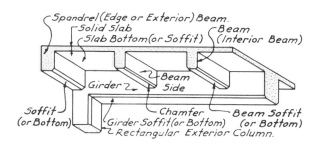

Fig. 13-2. A reinforced concrete, one-way floor system showing girder, beams and spandrel beam. (Concrete Reinforcing Steel Institute)

system is formed by using forms in a monolithic (one continuous) pour and tied directly to the column supports.

In the TWO-WAY FLOOR SYSTEM, the girders and beams (when used) run in two directions and are formed by steel "pans" laid over the form support structure. This system is sometimes referred to as a waffle slab, Fig. 13-3.

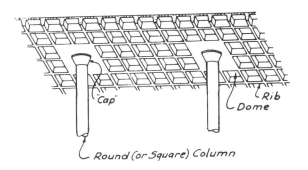

Fig. 13-3. Steel "pans" are used to form the support structure for this type of reinforced, two-way floor system. (Concrete Reinforcing Steel Institute)

Variations on the two-way system include:
a. Flat slab with beams running between columns.
b. Flat slab with flared columns and drop panels.
c. Flat plate construction with no beams or panels, Fig. 13-4, which is generally used for light commercial construction (office buildings and apartment houses).

Reinforcing steel is used in all of these floor systems. Its use is detailed on the blueprints for a specific location to provide maximum resistance to forces of compression, tension and shear. Reinforcing steel to be placed in concrete is designated with

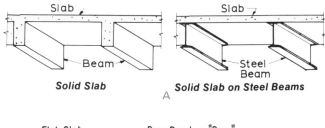

Solid Slab   Solid Slab on Steel Beams

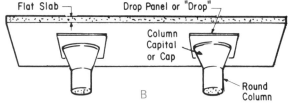

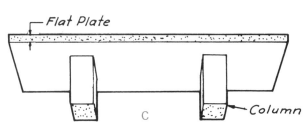

Fig. 13-4. Variations on the two-way floor system. (Concrete Reinforcing Steel Institute)

a note on the drawing giving the number of bars, size and placement:

2 - #5 x 19'-0" @ 2'-6" O.C.
PLACE ON 1 1/2" BAR CHAIR

The size of the bar is indicated by a number representing eighths of an inch. That is: a No. 5 bar is 5/8 inch. Identification marks rolled into the bars identify the producer's mill, bar size, type of steel and, for Grade 60, a grade mark indicating yield strength (Grade 40 and Grade 50 do not show a grade mark). See Fig. 13-5.

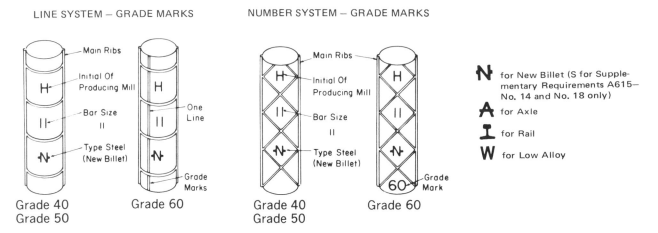

Fig. 13-5. Identification marks for steel reinforcing bars. (Concrete Reinforcing Steel Institute)

EXPANSION JOINTS of a flexible material usually are called for between the slab and wall or around columns, Fig. 13-6. These joints prevent the slab from cracking due to expansion and contraction, or they are used with ground level slabs that do not settle evenly with the footings and foundations walls. Other expansion joints may be planned in long slabs. These usually call for premolded joints and metal coverings.

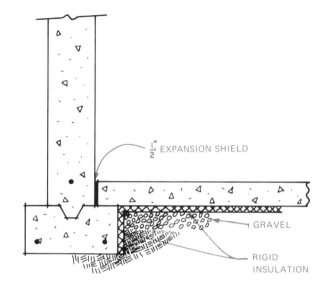

Fig. 13-6. *A flexible expansion joint material is placed between the concrete slab and wall.*

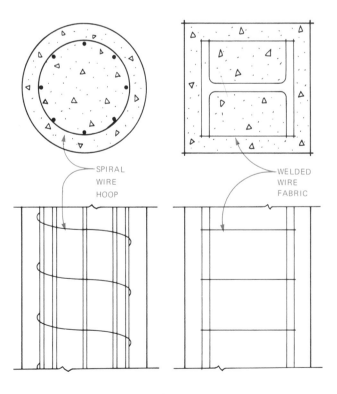

Fig. 13-7. *Two types of steel reinforced columns*

Some architects indicate on a drawing where it is permissible to make a joint when the pour is terminated at the end of the day. Where no indication is given, the location must be approved by the project engineer or architect.

COLUMNS are vertical support members designed to safely carry the anticipated final load of the building and its contents. Concrete columns are reinforced with steel rods or structural shapes, Fig. 13-7.

Columns may be dimensioned right on the drawing. Otherwise, they may be identified by a grid system of numbers and letters on the plan view of a drawing and called out on a column schedule, Fig. 13-8. Identification should include: size of the columns; number and size of vertical reinforcing rods; size and spacing of ties (horizontal reinforcing rods placed around the outside of the vertical rods).

COLUMN MARK	SIZE	BASE PLATE	SETTING PLATE	CAP. PLATE
A1				5" x 5/8" x 10"
A2				5" x 5/8" x 10"
A3				5" x 5/8" x 10"
B1				Thru Plate
B2				5 1/2" x 5/8" x 12"
B3				5 1/2" x 5/8" x 12"
C1	Typical 4" Ø Std. Col. x 10.79 #/Ft.	Typical 9" x 3/4" x 9" Base Plate	Typical 9" x 1/4" x 9" Setting Plate	Thru Plate
C2				5 1/2" x 5/8" x 12"
C3				5 1/2" x 5/8" x 12"
D1				Thru Plate
D2				5 1/2" x 5/8" x 12"
D3				5 1/2" x 5/8" x 12"
E1				5" x 5/8" x 10"
E2				5 1/2" x 5/8" x 12"
E3				5 1/2" x 5/8" x 12"

Fig. 13-8. *Schedule of columns.*

BEAMS are the structural members running horizontally beneath a floor system and anchored to columns, Fig. 13-9. Some beams are rectangular in cross section. Others are thinner at the bottom, where they are reinforced with steel to provide tension strength, and wider at the top to provide compression strength. See Fig. 13-10(a).

Beams are given added strength by stirrups (U-shaped steel rods) that hold the horizontal rods in place and increase the resistance to shear stresses, Fig. 13-10(b). Devices called "bolsters" and "chairs" are used to support reinforcing rods at the desired level off the bottom of the forms, Fig. 13-11.

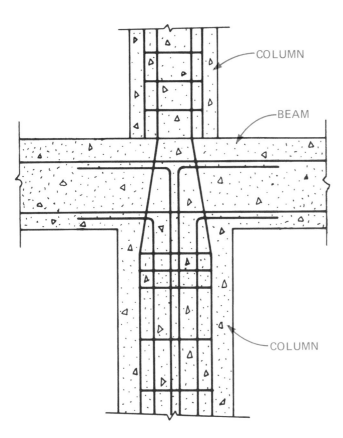

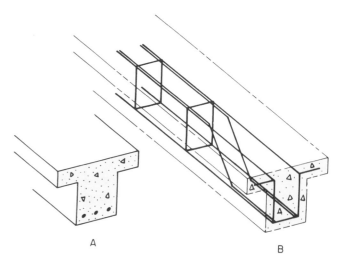

COLUMN

BEAM

COLUMN

Fig. 13-9. Beams are tied structurally into columns.

A

B

Fig. 13-10. Cross section of reinforced "T" beams.

A chart showing the standard types and sizes of bar supports and their symbols is included in the Reference Section.

Beams may be dimensioned on the drawing. However, if there are a number of beams, they will

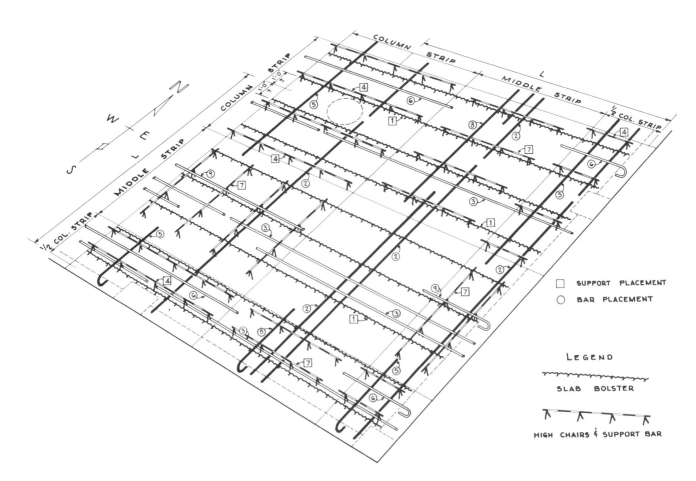

☐ SUPPORT PLACEMENT

◯ BAR PLACEMENT

LEGEND

SLAB BOLSTER

HIGH CHAIRS & SUPPORT BAR

Fig. 13-11. Bolsters and chairs hold the reinforcing steel at the right height in the forms. These are fastened to the forms and wired to the reinforcing rods. (Concrete Reinforcing Steel Institute)

be identified on the drawing and dimensioned in a schedule. Identification should include: beam size; number and size of reinforcing rods at the bottom and top of beam; number, size and type of stirrups.

SPLICING of reinforcing rods in columns and beams is necessary when the length of these members exceeds the length of the rods. Splicing is accomplished by overlapping the next rod by a length of so many times the diameter of the rod. A typical note controlling a lap would read:

30 DIA MINIMUM LAP

If No. 4 rod were specified, this would be 30 x 4/8 or 15 inches.

PIPING AND ELECTRICAL CONDUIT are placed in floor or roof slabs at the time reinforcing rods are installed, Fig. 13-12. Provision for pipes passing through a slab is accomplished by means of a fiber sleeve fastened to the decking and extending above the level of the concrete. These sleeves will be removed after the concrete dries and pipes will pass from floor to floor. Electrical conduits are

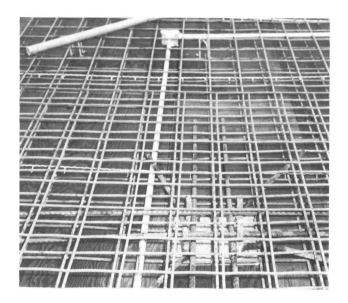

Fig. 13-12. Conduit is frequently placed in concrete floor slabs to serve electrical and telephone circuits. (Wire Reinforcement Institute)

installed with electrical boxes on the ceiling side of the slab, or conduit is stubbed up at a wall location or through a grid system to be located in the floor.

Structural Steel

STRUCTURAL STEEL refers to a system of building construction where the skeleton frame-

Fig. 13-13. Some structural steel shapes used in construction. (American Iron & Steel Institute)

work consists of structural steel shapes, principally wide flange (W) and "I" beams (S). See Fig. 13-13. The framework in heavy construction is designed so floor loads are transferred to columns of steel, so there is no need for load bearing walls. Steel framework members usually are encased in concrete or some other fireproofing material to protect them from failing during a fire.

Structural steel members are known as GIRDERS, BEAMS, SPANDREL BEAMS, COLUMNS and TRUSSES. See Fig. 13-14. Individual members may be dimensioned on the drawing or placed in a schedule if a sizable number of members are listed. A typical beam dimension would be W12 x 50, which means: a (W) wide flange shape; a nominal size of 12 inches in depth; a weight of 50 pounds per lineal foot.

STEEL JOISTS of the open web type have been widely used in light commercial construction for floor and roof structures. The Steel Joist Institute has standardized the specifications and labeling of these joists throughout the industry.

Steel joists may be grouped into three major series: SHORTSPAN, LONGSPAN and DEEP LONGSPAN JOISTS. The shortspan joists are the J-series and H-series. They are available in depths of 8 to 30 inches and spans up to 60 feet. J-series joists are based on design stress in tension of 22,000 psi minimum yield strength steel and H-series (high-strength) of 36,000 psi.

Longspan joists are labeled LJ and LH series. The L designates longspan. The LJ series has been standardized in depths from 18 to 48 inches and spans to 96 feet. Deep longspan joists have been standardized in depth from 52 to 72 inches for clear spans up to 144 feet.

A typical designation for a shortspan joist would be: 12J3, which means a shortspan joist of 12 inches in depth. The 3 is a range of span in feet (12 to 24 feet for this size joist), obtained from a load table to produce a structure of a certain load carrying capacity. A 60DLH15 is a deep longspan high-strength joist with a depth of 60 inches. From a load table, the 15 indicates a range of span from 70 to 120 feet, depending on the load required.

Steel joists are manufactured in both the TOP BEARING and BOTTOM BEARING types. Longspan and deep longspan joists, Fig. 13-15, come with parallel chords or pitched chords for roof drainage.

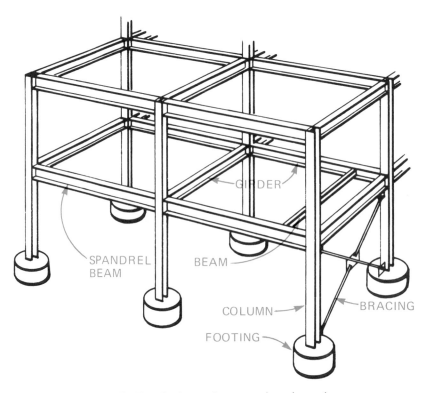

Fig. 13-14. Terminology of structural steel members.

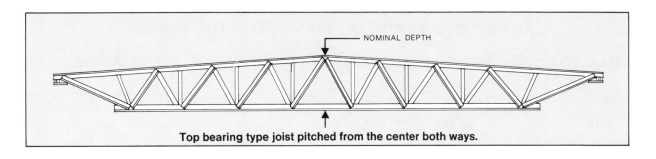

Top bearing type joist pitched from the center both ways.

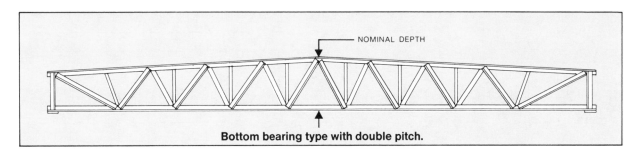

Bottom bearing type with double pitch.

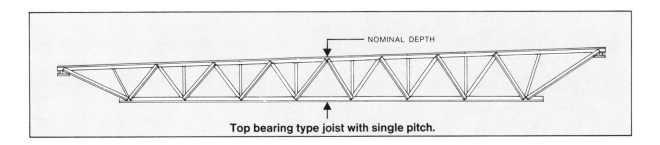

Top bearing type joist with single pitch.

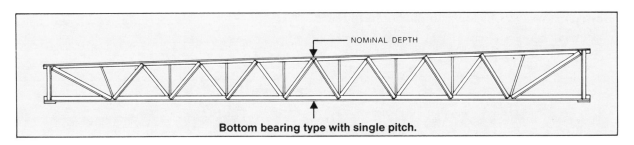

Bottom bearing type with single pitch.

Fig. 13-15. Types of open-web steel joists.   (Armco Steel Corporation)

Blueprint Reading Activity 13—1
REINFORCED CONCRETE AND STRUCTURAL STEEL FRAMING

Refer to Prints 13-1a and 13-1b in the Large Prints Folder to answer the following questions.

Print 13-1a

1. What size steel girders and beams are used to frame the roof of the Porte-cochere? Interpret the specification sizes.

1. _____

2. How are the cross beams attached to the girders?

2. _____

3. Describe the column at A and its reinforcement.

3. _____

4. How is the beam at A held to column?

4. _____

5. Describe how the same beam is anchored to the column at B.

5. _____

6. What structural roof material is placed on top of these steel beams?

6. _____

Print 13-1b

7. What elevation level is assigned to the first floor? Fourth floor?

7. _____

8. What size and how many vertical reinforcing bars are specified for between the second and third floors?

8. _____

9. How thick is the wall at C and what is the minimum distance permitted between the vertical bar and the opposite wall exterior face?

9. _____

10. What size and at what spacing are the vertical bars between:
    a. Third and fourth floors?
    b. Fourth floor and roof?
    c. What is to be done with the bars at the roof level?

10. a. _____
    b. _____
    c. _____
    _____

11. How many, what size and what spacing is to be used on the vertical bar ties at the end corbels between the second and third floors?

11. _____
    _____

12. How many bond beams are called for between the second and third floors? What reinforcement is required?

12. _____
    _____

13. What is the size of the dowel at D anchoring the fourth floor to the wall and how are they to be placed?

13. _____
    _____
    _____

Refer to Prints 13-2a and 13-2b in the Large Prints Folder to answer the following questions.

Print 13-2a

1. Interpret the information given for the column footing at A.

1. _____

2. Give the top and bottom elevations of the footing at B.

2. Top: _____
   Bottom: _____

3. Describe the reinforcement around openings in the foundation wall.

3. _____

4. What reinforcement is required in the foundation wall at Section 3-3?

4. _____

5. How is each door jamb reinforced?

5. _____

6. Describe the reinforcement above door openings in Section 6-6.

6. _____

7. What specification is given for the concrete?

7. _____

8. Give the specifications for the reinforcing steel.

8. _____

9. How is the floor joist at C to be secured to the foundation wall?

9. _____

Print 13-2b

10. Give the steel beam size at D and explain the sizing.

10. _____

11. Give the size of the beam at E and explain the sizing.

11. _____

12. How is the beam secured to the masonry wall at F?

12. _____

13. Give the size of the roof joist at G and explain the sizing.

13. _____
_____
_____

14. How far over-centers are the joists at G?

14. _____

15. What size, type and spacing of floor joists are required at H?

15. _____
_____
_____
_____

Unit 14
Framing and Finish Construction Blueprints

Framing of a building entails all of the work of wood and/or metal that is involved in assembling the structure. Framing gives the building shape and strength. It includes the framing members of floors, exterior and interior walls, ceiling and roof, along with the subflooring, wall and roof sheathing.

Blueprints usually do not detail basic construction procedures. Framing plans, however, do detail SPECIAL CONSTRUCTION PROCEDURES, and it is important that workers interpret these in the manner specified on the drawing.

This unit is designed to assist you in reading and interpreting framing and exterior and interior finishing blueprints.

Floor Plans

Floor plans provide more information on the construction project than any of the other drawings. The sizes of the various rooms and the traffic patterns in a house or commercial building affect to a large extent the exterior shape and size of the structure.

Floor Framing Terminology

Study the illustrations in Fig. 14-1 to become familiar with the terms used for various framing members. Note that the sill rests on the foundation wall, and the header and floor joists rest on the sill. Other floor framing members are identified in Fig.

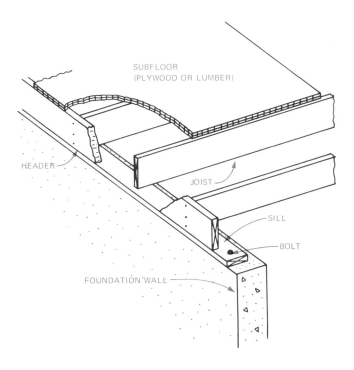

Fig. 14-1. Floor framing terminology (platform construction).

14-2. Wherever openings cut across floor joists, it is standard practice to use two joists (double trimmer) and two headers (double header) on each side of the opening.

Details on the drawing are used to clarify construction procedures for a particular feature of the building. Often, these details call out the size of framing members to be used, Fig. 14-3.

Dimensions for exterior walls usually are given to the outside of the stud wall for frame and brick

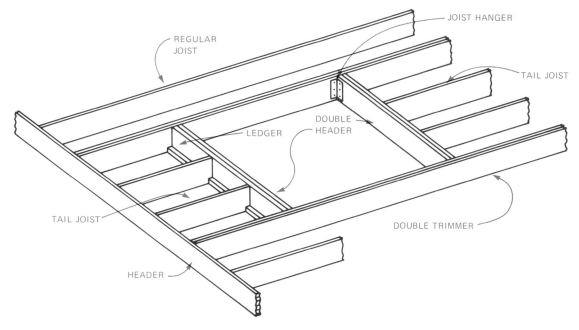

Fig. 14-2. Additional members used in floor framing.

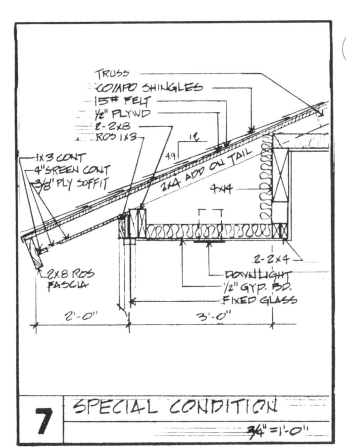

Fig. 14-3. Detail drawing of a framing procedure calling out size of members. (Schweizer Associates Architects, Inc.)

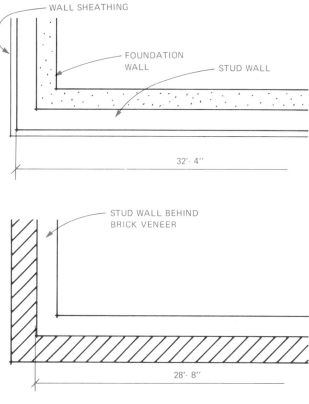

Fig. 14-4. Dimensions shown are to face of the stud wall.

veneer buildings. See Fig. 14-4. As noted in Unit 8, some architects follow the practice of starting the dimensions for single-story frame buildings at the surface of the wall sheathing (which should align with the foundation wall). A note may be added to the drawing to read:

NOTE: EXTERIOR DIMENSIONS ARE TO OUTSIDE EDGE OF STUDS;

INTERIOR DIMENSIONS ARE TO CENTER OF STUDS

The drawings should be checked carefully to verify the dimensioning practice followed.

Interior walls of frame construction usually are dimensioned to their center lines. Masonry interior walls are dimensioned to their faces with the wall thickness also dimensioned. (See Figs. 8-7 and 8-8)

In studying the floor plans of buildings with two or more stories (floors), note particularly the adjoining stairs, chimneys and load-bearing partition walls. In certain style buildings, the floor is cantilevered out over the foundation wall. In these cases, special framing details usually are shown for such floor extensions. See Fig. 14-5. Houses having second stories smaller than the first are called one-and-a-half story houses. These houses usually involve "knee walls" (a short wall joined by a sloping ceiling) and dormers.

Split-level houses have floor plans in which the levels, or stories, are separated by half-flight of stairs. Many variations are called for in framing, so the plans should be studied carefully.

It is important to know that notes on a floor plan, referring to the framing of joists, relate to the level above. For example, a note on the foundation plan, such as

$$\overline{\underline{\text{2 X 10 JOISTS}}\atop 16'' \text{ O.C.}}$$

refers to the first-floor framing overhead. In the same manner, a similar note on the first-floor plan of a building refers to the framing of the joists overhead.

Wall Framing Drawings

There are three basic types of light frame construction: platform; balloon; plank and beam. The construction worker should be familiar with the three types and be able to distinguish between them on drawings.

PLATFORM FRAMING, also know as Western Framing, is the type most widely used. It gets its name from the appearance it presents in framing. The first floor is built on top of the foundation, so

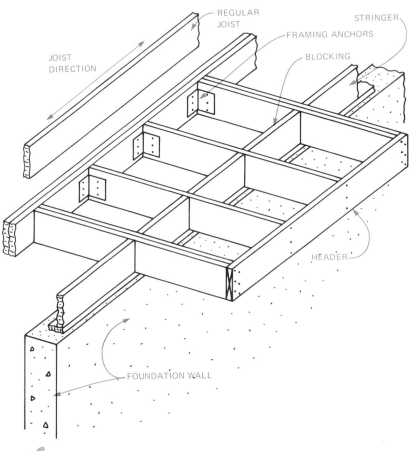

Fig. 14-5. Pictorial detail showing framing of a cantilevered floor.

it appears like a "platform" when the subflooring is complete. The first-floor wall sections are raised and a second-floor "platform" is built on top of these walls. Then, the second-floor wall sections are raised and another "platform" for the second story ceiling or floor of the next level is constructed. See Fig. 14-6. Each floor is a separate unit built on the structure below.

BALLOON FRAMING is not used to any large extent today. This type of framing is characterized by the studs extending from the first floor sill to the top floor plate, as in Fig. 14-7. Second floor joists rest on a member called a "ribbon" set into the studs. Balloon framing has some advantages: it reduces lumber shrinkage problems in masonry veneer or stucco structures. It simplifies running ducts and electrical conduit from floor to floor.

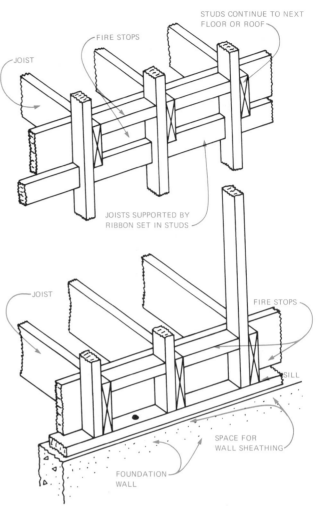

Fig. 14-7. Details of balloon framing.

The main disadvantage of balloon framing is in the tendency of the walls to act as flues in spreading fires from floor to floor unless blocking is added. This type of framing also is more difficult to manage in assemblying a wall section.

PLANK AND BEAM FRAMING consists of heavy timber material for posts in wall sections and beams of two inch plank material supporting floor and roof sections. See Fig. 14-8. The structural members are placed at wider intervals than in the other methods of framing. This type of framing lends itself to interesting architectural effects and extensive use of glass and exposed wood sections, as shown in Fig. 14-9.

WALL FRAMING TERMINOLOGY is illustrated in Fig. 14-10. Note that the frame wall starts with the sole and ends with the double plate. In one-story buildings, the ceiling joists and roof

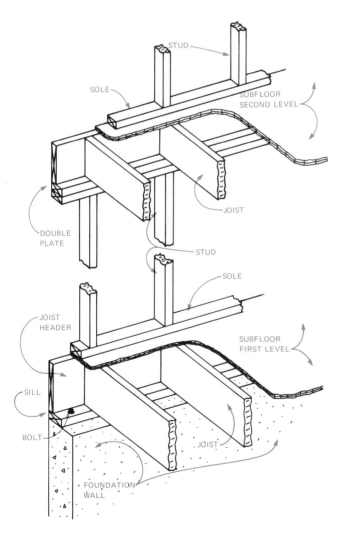

Fig. 14-6. An example of platform framing.

Fig. 14-8. Plank and beam framing using transverse beams. Longitudinal beams would run parallel to the ridge beam. (Brown & Kauffman Architects)

Fig. 14-9. A residence making use of the plank and beam system of framing. (Southern Forest Products Association)

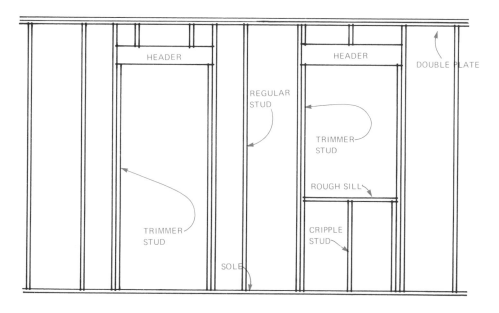

Fig. 14-10. Terminology of a frame wall.

rafters rest on this double plate. In multi-story buildings using platform type construction, the next floor is built on the double plate. Headers carry the weight of ceiling and roof across framing openings for doors and windows. The dimensions of headers and rough opening are shown on elevation framing or sectional views. Window and door openings have a regular stud plus a trimmer stud to give strength and rigidity to the opening.

INTERIOR WALLS that carry the ceiling or floor load from above are called "bearing partitions." Usually, they are located over a beam or bearing partition wall below, Fig. 14-11.

DOOR AND WINDOW SCHEDULES give full information on the number and sizes of the various doors and windows in the building. See Fig. 14-12. Units listed in the schedule are referenced to the plan view with a letter or number. Some architects provide the rough opening size in the schedules to

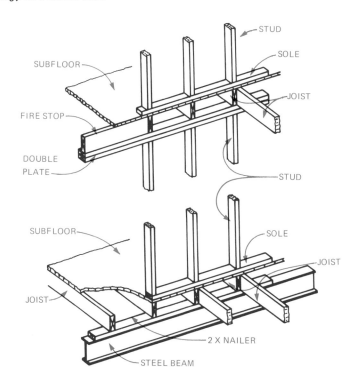

Fig. 14-11. Bearing partitions are used to transfer loads from floors above to a bearing structure below.

DOOR SCHEDULE

MARK	TYPE	SIZE	MATERIAL	FRAME	REMARKS
A	1	3'-0" x 7'-0" x 1 3/4"	Hollow Metal	Holl. Met.	Closer and Threshold.
B	1	3'-0" x 7'-0" x 1 3/4"	Hollow Metal	Holl. Met.	Closer
C	2	2'-8" x 7'-0" x 1 3/4"	Hollow Metal	Holl. Met.	Closer and Kick Pl.
D	2	3'-0" x 7'-0" x 1 3/4"	Hollow Metal	Holl. Met.	Closer
E	2	3'-0" x 7'-0" x 1 3/4"	Hollow Metal	Holl. Met.	Closer

Fig. 14-12. A typical door schedule.   (Smith & Neubek & Associates)

speed construction and assure a correct fit. If rough opening sizes for doors and windows are not provided, the construction worker must calculate these in framing wall openings.

SECTIONAL VIEWS OF WALLS usually are drawn to a larger scale and included on the drawings to clarify construction details, Fig. 14-13. These sections are referenced to the plan view with a reference line designating the location of the wall section in the building. (See Fig. 4-1.) NOTE: The section is viewed in the direction of the arrows.

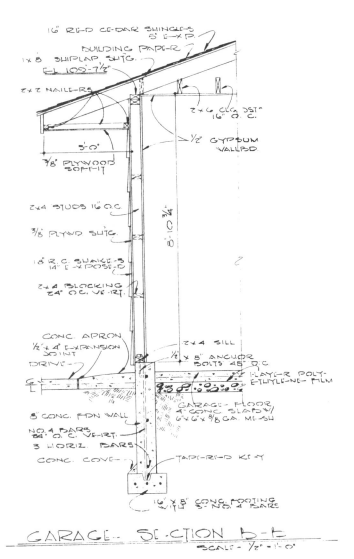

Fig. 14-13. Wall section details provide necessary framing information. (Garlinghouse Plan Service)

TRANSVERSE AND LONGITUDINAL SECTIONS are full sections cut through the width or length of a building, Fig. 14-14. They are prepared for buildings with more complex framing problems such as split-level houses or those having unusual interiors. Transverse sections are taken through the narrow dimension and the longitudinal section is through the length of the building. These sectional views show features such as floors, walls and ceiling as sections. Features beyond the cutting plane are shown as they appear in the interior of a building.

Ceiling Framing Drawings

Ceiling framing is similar to floor framing and, except for the ceiling immediately below the roof, the floor joists carry the ceiling on multi-story buildings. Ceiling joists for flat roofs are sometimes the same as the roof joists or rafters. Ceiling joists which serve as support for ceilings only are lighter members and do not include headers around the outside, Fig. 14-15.

Roof Framing

Construction workers should be familiar with the different types of roofs, and how they are framed. Sketches of various roof styles found in house construction are shown in Fig. 14-16. The style of roof is most easily identified in elevation drawings. Some architects do not supply a roof framing drawing for the more common type roofs, such as the gable or flat roof, but depend on the elevation and detail drawings to guide the construction workers. When the roof is more complicated, or when the architect desires to specify the manner of construction, a roof framing plan is prepared, Fig. 14-17.

ROOF FRAMING TERMINOLOGY is illustrated in Fig. 14-18. In the plan view, Fig. 14-19, several roof types and kinds of rafters are shown. COMMON RAFTERS run at right angles from the wall plate to the ridge. HIP RAFTERS extend from an outside corner of the building to the ridge board, usually at a 45 degree angle. VALLEY RAFTERS extend from an inside corner of the building to the ridge board, usually at a 45 degree angle. JACK RAFTERS are those that do not extend to the ridge board: hip jack from plate to hip rafter; valley jack from plate to valley rafter; cripple jack between valley and hip rafters.

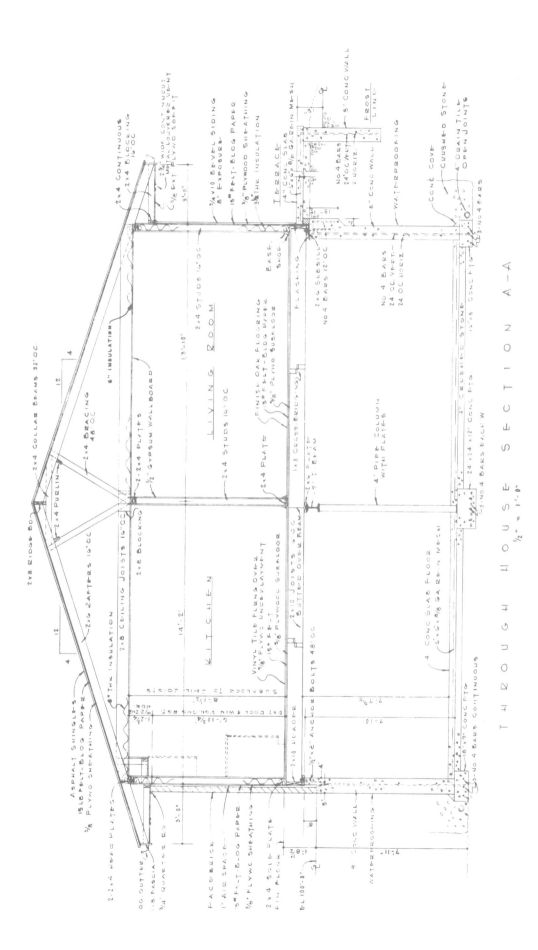

Fig. 14-14. Transverse and longitudinal sections help clarify construction details. (Garlinghouse Plan Service)

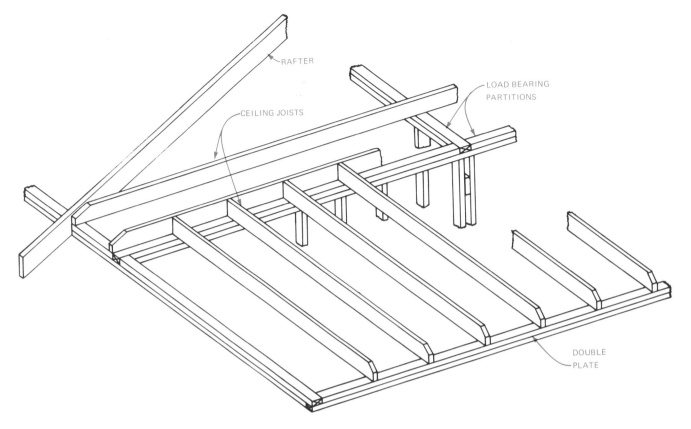

Fig. 14-15. Ceiling joists are fastened to the double plate.

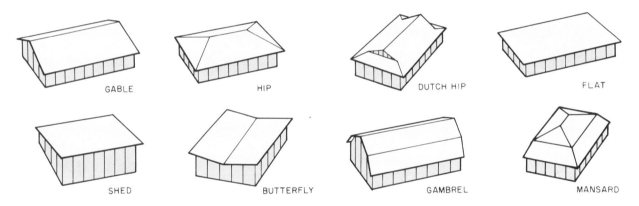

GABLE

HIP

DUTCH HIP

FLAT

SHED

BUTTERFLY

GAMBREL

MANSARD

Fig. 14-16. Common roof styles for residences.

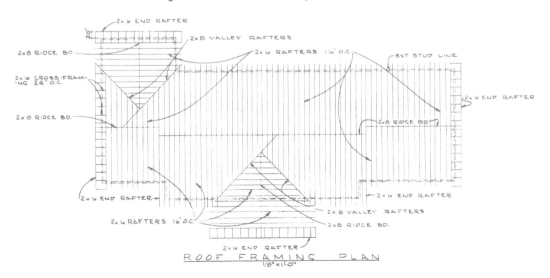

ROOF FRAMING PLAN
1/8" = 1'-0"

Fig. 14-17. A typical roof framing plan for a residence.

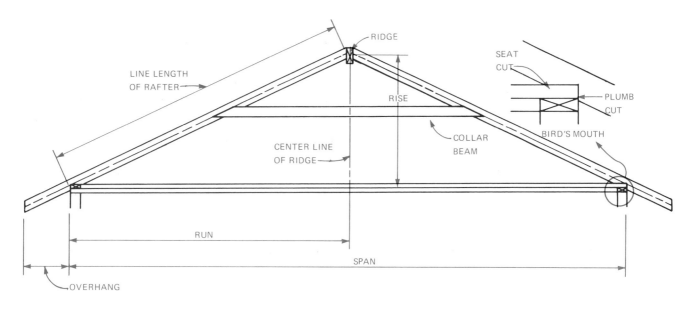

Fig. 14-18. Roof framing terminology.

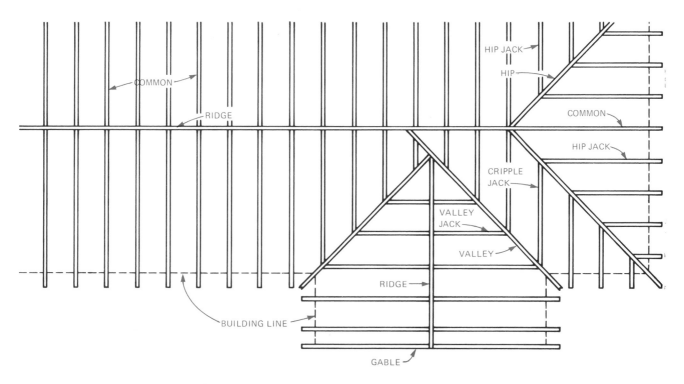

Fig. 14-19. Identification of various kinds of rafters.

Another roof framing member is the PURLIN, a horizontal member laid over a truss to support long rafters or the break in gambrel roof rafters.

Stair Framing

STAIRWAYS, Fig. 14-20, are of three general types: (A) straight run; (B) long L; (C) double L and U. A modification of the long L features "winders" or "pie-shaped" treads instead of the landing as a space saver. Also, there are circular winding stairs that gradually change direction as they ascend or descend. STAIR TERMINOLOGY is shown in view (A) in Fig. 14-20. The number of risers in a stair is always one greater than the number of treads. The stringer or carriage usually is cut and installed at the time of floor and wall framing. "Rough" treads are installed for use during construction. These are replaced at the time finish work is performed.

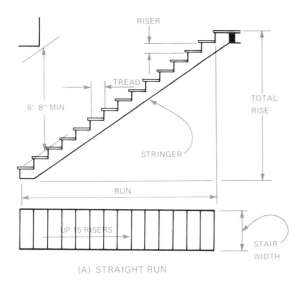

(A) STRAIGHT RUN

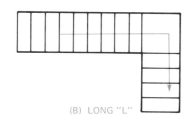

(B) LONG "L"

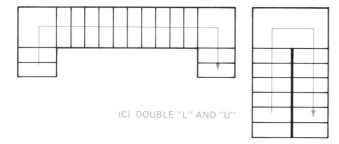

(C) DOUBLE "L" AND "U"

Fig. 14-20. Types of stairways and stair terminology.

Metal Frame Construction

A number of manufacturers are producing steel and aluminum framing systems for commercial applications. These framing members are manufactured as individual items to be assembled later on the job site or as preassembled components. The floor framing system is fastened to the foundation with powder-set fasteners. Plywood subflooring is fastened to the metal joists with power-driven screws or nails. Gypsum wallboard or other wall covering material is fastened to the studs with power-driven screws. A note on the drawing will indicate the type of metal framing used.

Exterior/Interior Finish

The elevation drawings and wall sections show the exterior finish. This includes siding, cornice, roofing, exterior window and door designs, Fig. 14-21. Also, dimensions of ceiling heights, roof pitch and cornice details are given. Window units are designated by an "O" for a fixed section and by an "X" for a sliding section.

Interior finish information is placed in tabular form on the drawing in a FINISH SCHEDULE.

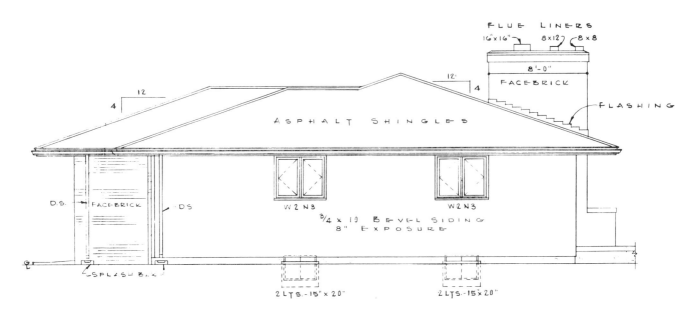

Fig. 14-21. Exterior siding material and finishes are specified on elevation drawings.   (Garlinghouse Plan Service)

Blueprint Reading Activity 14—1
WOOD FRAMING AND FINISH CONSTRUCTION BLUEPRINTS

Refer to Prints 14-1a through 14-1d in the Large Prints Folder to answer the following questions.

1. What is the scale of:
   a. The floor plan?
   b. The framing drawings?

   1. a. _____
      b. _____

2. What size are the pipe columns supporting the I-Beam? How many are there?

   2. _____

3. How many I-Beams are there, and what size are they?

   3. _____
      _____

4. Give the size of:
   a. The plate over the I-Beam
   b. The sill on the foundation wall

   4. a. _____
      b. _____

5. Give the following dimensions:
   a. Overall length to face of studs
   b. Overall width to face of studs

   5. a. _____
      b. _____

6. What is the size of the floor joists, and how are they laid over the beam?

   6. _____

7. Describe the floor joist construction under the partitions running parallel to the floor joists.

   7. _____
      _____

8. For the fireplace floor framing:
   a. Give opening size and location

   8. a. _____
         _____
         _____

   b. Describe framing of opening

      b. _____
         _____

9. Describe the bridging of the floor joists, including the blocking on the ends of the house.

   9. _____
      _____
      _____

10. Give the dimensions for A, B and C.

    10. A _____
        B _____
        C _____

11. What is the width of the foyer to face of studs? How far is the stoop recessed from the front stud face?

11. _____
_____

12. Give the distance from corner D to the center of rear door.

12. _____

13. Describe the floor framing for the stairway:
    a. Opening
    b. Carriages
    c. Headers
    d. Kickplate
    e. Risers
    f. Treads

13. a. _____
    b. _____
    c. _____
    d. _____
    e. _____
    f. _____

14. What does Section X-X show?

14. _____
_____

15. What type of Section is A-A?

15. _____

16. Give the size of the studs in the regular walls and their spacing.

16. _____

17. What are the header sizes over:
    a. The garage door
    b. The front windows
    c. The front stoop

17. a. _____
    b. _____
    c. _____

18. How many jamb studs are used to support each end of the garage door header?

18. _____

19. What size beam is used at E?

19. _____

20. Give the rough opening height for the garage doors.

20. _____

21. What size beam is used in the garage to support the ceiling joists?

21. _____

22. What material is used for:
    a. Roof sheathing?
    b. Roofing?

22. a. _____
    b. _____

23. What size ceiling joists are used at F and G?

23. F _____

    G _____

24. Describe the ceiling framing above the wall at H. See detail drawing, Sheet 4.

24. _____

    _____

25. What size is the roof ridge board?

25. _____

26. Give the size of:
    a. Common rafters
    b. Valley rafters
    c. Collar beams
    d. Purlin

26. a. _____

    b. _____

    c. _____

    d. _____

Blueprint Reading Activity 14—2
METAL FRAMING BLUEPRINTS

Refer to Prints 14-2a through 14-2f in the Large Prints Folder to answer the following questions.

1. What is the scale of:
   a. Lower floor plan
   b. Second floor framing plan
   c. Typical column and footing detail

1. a. _____
   b. _____
   c. _____

2. How many columns and what size are used to support the second floor and roof beams?

2. _____

3. What is the overall size of the foundation plan?

3. _____

4. Give the size of:
   a. Second floor beam between the columns.
   b. Second floor beam over entry doors.
   c. Roof beam center portion of building.
   d. Roof beam at flat portion of roof.

4. a. _____
   b. _____
   c. _____
   d. _____

5. How are the beams to be fastened to the columns?

5. _____

6. How is the roof beam to be secured to the wall?

6. _____
   _____
   _____
   _____
   _____

7. What is the distance between the finished floor elevations of the first and second floors?

7. _____

8. How many and what size joists are required for the second floor framing?
   a. Center section
   b. End sections

8. a. _____
   b. _____

9. What bridging and bracing are required for the above joists?

9. _____
   _____

10. How many risers are required in the stairway?
    a. Between the entrance floor level and the stair landing?
    b. Stair landing to second floor?

10. a. _____
    b. _____

11. Give the tread size and run of the above stairway from:
    a. Entrance floor level to stair landing
    b. Stair landing to second floor

11. a. _____
    b. _____

12. How thick is the concrete on the stair landing?   12. _____

13. What material is used for the stair handrail, and how is it hung?   13. _____ _____

14. What is the size of the stair landing?   14. _____

15. Give the size of the following for the wall construction:
    a. Steel studs
    b. Drywall
    c. Soffit plywood

    15. a. _____
        b. _____
        c. _____

16. How is the masonry wall to be treated under each second floor joist?   16. _____ _____

17. What is the material used for the second floor?   17. _____ _____ _____

18. Give the number and size of the roof joists required for:
    a. Center section
    b. End sections

    18. a. _____
        b. _____

19. How are the roof joists spaced at:
    a. Front center section
    b. End sections

    19. a. _____
        b. _____

20. What materials are used above the roof joists to form the roof?   20. _____ _____ _____ _____

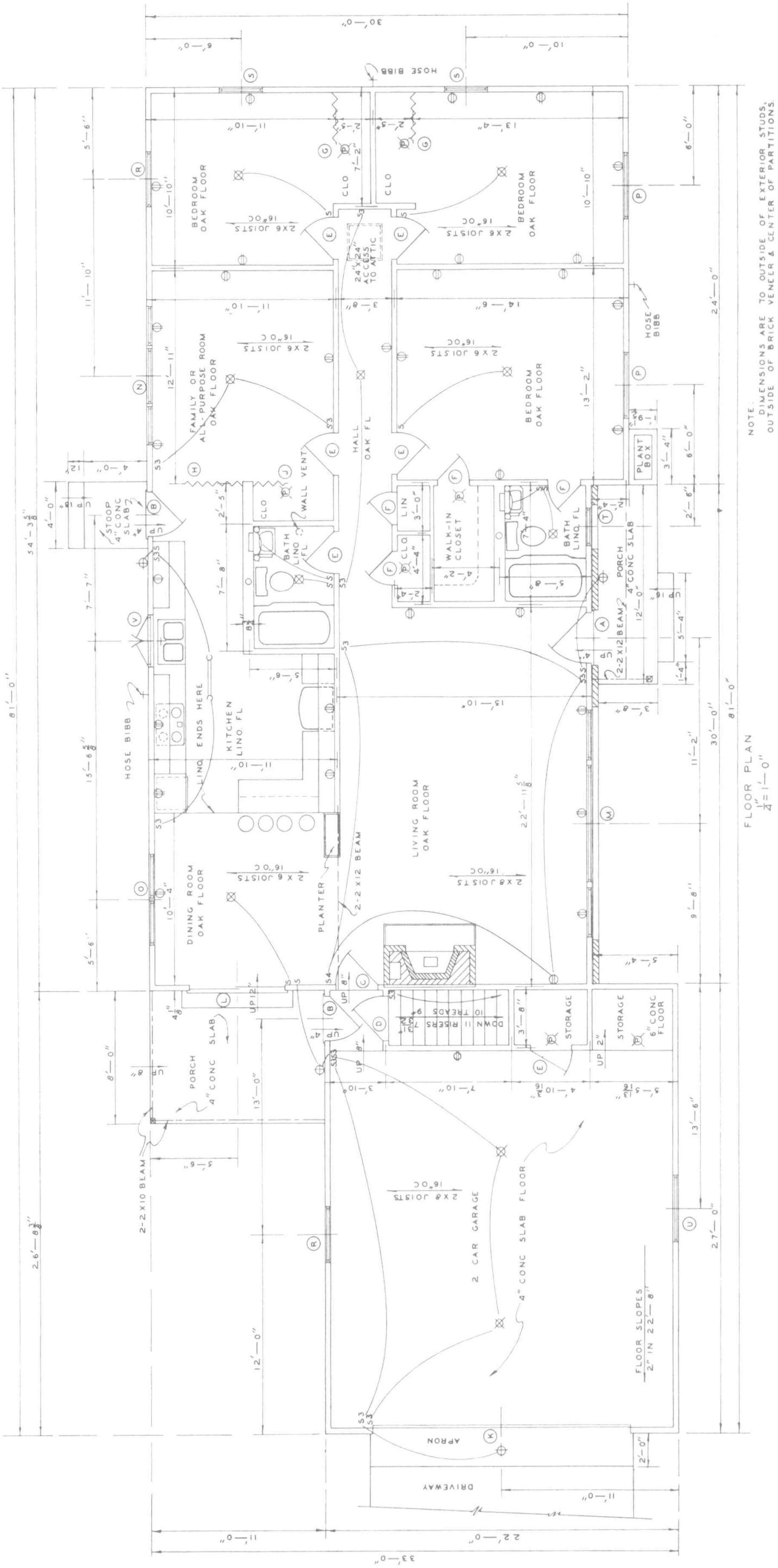

Fig. 15-1. A floor plan showing location of plumbing fixtures.

Unit 15
Plumbing System Blueprints

Plumbing in most residences and light commercial construction consists of the water distribution system, sewage disposal system and some plumbing in connection with the heating and cooling systems. This unit is designed to familiarize you with the way in which these systems are detailed on the various construction blueprints.

Actually, piping diagrams are seldom prepared for residences. Symbols are shown on the plan drawings for location of fixtures such as sinks, water closets, floor drains and exterior hose bibbs. See Fig. 15-1. The plumber installs the system in accordance with the specifications and local government codes. Larger construction jobs have complete plumbing system drawings showing the supply and sewage piping.

Specifications

The specifications which relate to the plumbing should be read carefully before the job is started. These specifications detail work to be done in the plumbing part of the contract, piping materials to be used, type and quality of fixtures, and tests to be performed on the water distribution and sewage disposal system.

Plumbers must coordinate their work assignments with other tradesworkers since plumbing takes place at three different stages of construction:
1. Provisions for a water outlet and for the service entrance of the water supply and sewer drain to the building usually are made prior to the pouring of the foundation.
2. The next stage is the "rough-in" phase of the plumbing, which includes the running of water supply pipes and sewage drain pipes. The "rough-in" work is performed before the slab is poured in slab-in-ground construction and before wall covering materials are placed on the wall framing. See Fig. 15-2.
3. The final stage of the plumber's work is the finish plumbing, which includes the setting of fixtures after the walls and floors are finished.

Fig. 15-2. A rough-in of plumbing in a large apartment building. (Cooper Development Association, Inc.)

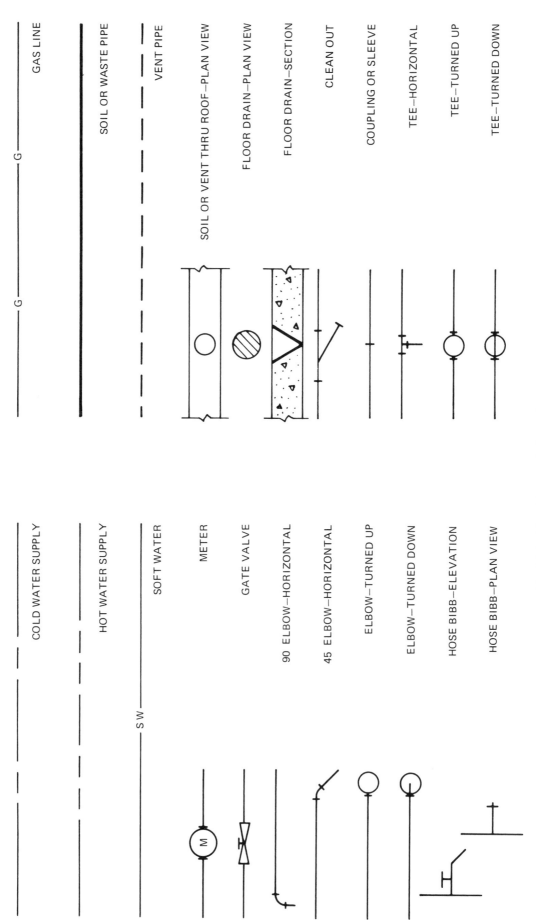

GAS LINE

SOIL OR WASTE PIPE

VENT PIPE

SOIL OR VENT THRU ROOF—PLAN VIEW

FLOOR DRAIN—PLAN VIEW

FLOOR DRAIN—SECTION

CLEAN OUT

COUPLING OR SLEEVE

TEE—HORIZONTAL

TEE—TURNED UP

TEE—TURNED DOWN

COLD WATER SUPPLY

HOT WATER SUPPLY

SOFT WATER

METER

GATE VALVE

90 ELBOW—HORIZONTAL

45 ELBOW—HORIZONTAL

ELBOW—TURNED UP

ELBOW—TURNED DOWN

HOSE BIBB—ELEVATION

HOSE BIBB—PLAN VIEW

Fig. 15-3. Plumbing symbols used on drawings.

Water Distribution System

The water distribution system includes the "main" supply line from the municipal water meter (or other source of supply) to the building. All pipes which take the water from the main to the various service outlets (water heaters, sinks, water closets, hose bibbs, etc.) are called distribution pipes. The distribution system also includes all control valves within the system.

The floor plan shows the location of plumbing fixtures as well as floor drains and exterior hose bibbs, Fig. 15-1. Symbols used to represent plumbing fixtures are pictorial representations of the fixtures on the plan views and on interior elevation views. The most common symbols used for plumbing are shown in Fig. 15-3. Unless the piping layouts are unusual, they generally are not shown on residential drawings. They are included on commercial projects.

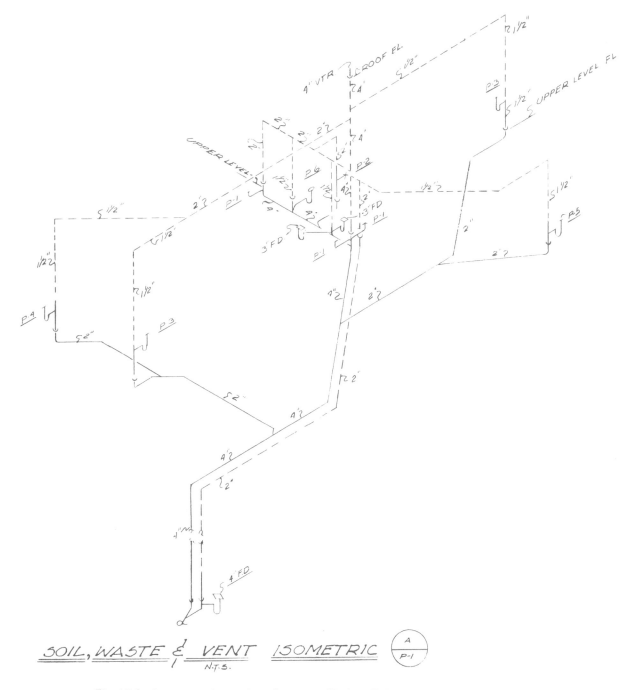

Fig. 15-4. A waste and vent riser diagram.   (Robert E. Hayes & Associates Architects)

PIPING MATERIALS for water distribution includes copper, which comes in heavy, medium and light wall thicknesses. The heavy and medium pipes are suitable for underground and interior plumbing systems. Light wall pipe is suitable for interior plumbing applications.

COPPER PIPING is available in hard and soft tempers. The hard tempers are more suitable for straight runs for exposed piping, where neatness in appearance is a factor. The hard temper pipe can be bent to a limited radius with proper tools. Soft temper copper pipe is easily bent by hand or with a tube bender.

Copper piping should not be embedded in concrete slabs, masonry walls or footings. When it is necessary for the piping to go through a slab or wall, a sleeve of asbestos or a larger pipe should be placed between the copper water piping and concrete to permit movement due to expansion of the copper. Copper has many advantages and is widely used in water distribution systems.

GALVANIZED STEEL PIPE has great strength and dimensional stability. The galvanized coating protects the inside and outside of the pipe. There are three grades of wall thicknesses of galvanized steel pipe: standard weight, extra strong and double extra strong pipe. The standard weight pipe is used for most residential and light commercial water distribution systems.

BRASS PIPE is used where the water is highly corrosive, such as coastal areas where salt water is used for cooling, baths or other applications.

PLASTIC PIPE is gaining in acceptance in some areas for water transmission systems and, more particularily, for sprinkling systems. Three of the most common types are acrylonitrile butadiene styrene (ABS), polyvinyl chloride (PVC) and polyethylene. There are threaded plastic pipes and fittings, but the most common plastic pipes and fittings have solvent-welded joints.

Sewage Disposal System

The SEWAGE DISPOSAL SYSTEM includes a vertical soil (waste) stack, a vent and a trap for each fixture. The waste stack carries the water materials to the building drain, to the building sewer line outside the building and on to the public sewer or septic tank. At the base of each stack, fittings called "cleanouts" (CO) are placed so the drain lines may be cleaned out with a plumber's rod or tape. The installation of sewage disposal systems is carefully controlled by plumbing codes to prevent contamination of the water supply and to keep sewer gases from entering the building.

PIPING DRAWINGS OR SCHEMATICS are provided for most commercial construction projects. Fig. 15-4 shows a waste and vent riser diagram for a light commercial building from the building drain to the vent through the roof (VTR).

A plan view diagram of a water distribution system and a sewage disposal system is shown in Fig. 15-5. Note the size of drain pipe specified. Vertical soil pipe or waste stacks serving water closets must have a thicker wall with at least six inch studs to accommodate a four inch soil pipe and its joints. In some cases, an isometric drawing is provided to detail the layout of the sewage disposal system.

Materials: Sewage disposal systems of cast-iron pipe have good qualities of strength and resistance to corrosion. Copper and plastic pipe have been gaining in use in recent years because of ease of installation. Lead pipe is used for drainage systems where chemicals are used. Clay tile and plastics (ABS and PVC) are used extensively for sewer piping.

Gas and Fuel Oil Systems

Sometimes the piping for gas or oil fired heating systems are included in the plumbing contract. Other than the location of the particular service desired, piping for heating purposes usually is not shown on the drawing. Where piping layouts are given, symbols are included. See Fig. 15-3. Specifications for the construction job will detail the size and kind of piping to use.

Materials most commonly used for gas piping are black wrought iron, galvanized steel or yellow brass. Copper tubing is banned by most building codes because it is corroded by most gases.

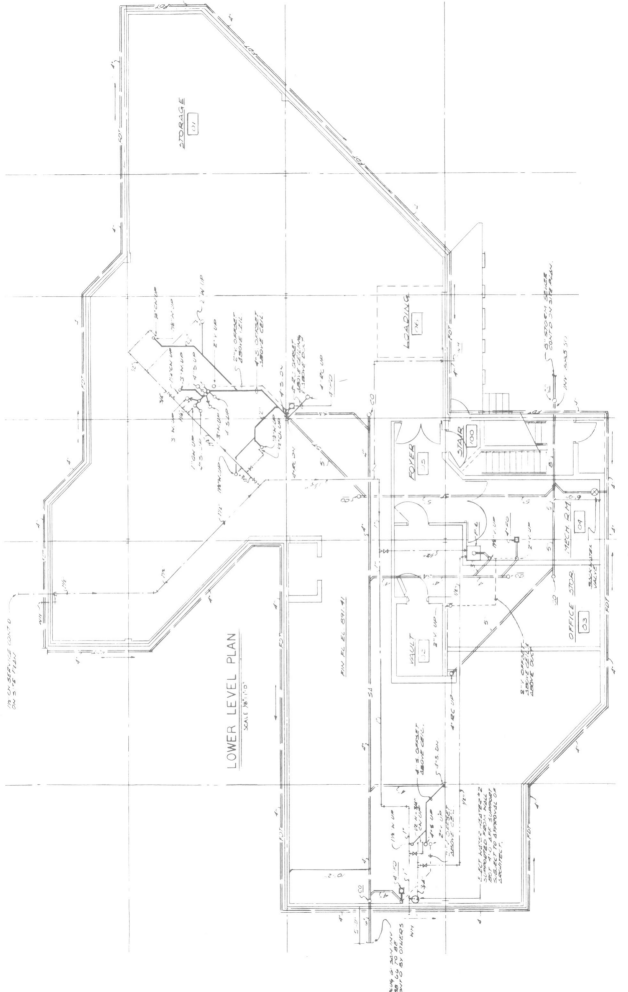

Fig. 15-5. Plumbing diagram of a water distribution system and a sewage disposal system. (Robert E. Hayes & Associates Architects)

147

Plumbing Codes

The Uniform Plumbing Code and the local government code control all aspects of the plumbing work, including the kinds and sizes of pipe, location of traps and cleanouts, plumbing fixture requirements, venting provisions and connections to water supply and to sewer lines. These codes also specify the tests for leaks which must be conducted on water supply lines and waste lines.

Blueprint Reading Activity 15—1
PLUMBING SYSTEM BLUEPRINTS FOR A RESIDENCE

Refer to Prints 15-1a and 15-1b in the Large Prints Folder to answer the following questions.

1. What is the size of the building sewer?

1. _____

2. Where does the sewer lead?

2. _____

3. What is the scale of the plumbing diagram?

3. _____

4. What part of the house does the plumbing riser diagram A detail?

4. _____

5. What size waste pipe is used to drain the kitchen sink and dishwasher?

5. _____

6. What size is the vent through the roof near the lavatory in the master bath?

6. _____

7. In the same bathroom, what is the size of the VTR, and where is it run?

7. _____

8. The diagram indicates what size waste stack is to be used for water closet in the master bath?

8. _____

9. What size vent through the roof and waste pipe is to be used for the clothes washer on the second floor?

9. _____

10. The 3" vent through the roof in diagram B serves what fixtures?

10. _____
_____

11. How many sewer "cleanouts" are indicated on the plumbing riser diagram?

11. _____

12. What type of interior water piping is specified?

12. _____

13. How shall this interior copper supply pipe be protected? Explain:

13. _____
_____
_____
_____
_____
_____
_____
_____
_____

14. How many and what brand and model of water closets are specified?

14. _____
_____

15. What plumbing valves are specified?

15. _____

16. What model kitchen sink is specified?   16. _____

17. What must the plumbing contractor provide regarding the air conditioning unit?   17. _____

18. How many hose bibbs are shown on the plans?   18. _____

19. Who is responsible for providing temporary electrical, water and sanitary facilities during construction?   19. _____

20. Plumbing permits are whose responsibility?   20. _____

Refer to Prints 15-2a and 15-2b in the Large Prints Folder to answer the following questions.

1. What is the size of the main supply pipe from the meter?

1. _____

2. What is the size of the main pipe as it enters the building?

2. _____

3. List the size of the distribution system pipe:
   a. To electric water heater
   b. From electric water heater
   c. To service sink
   d. Lavatories
   e. Water closets
   f. Dwyer kitchen unit
   g. Electric drinking fountain
   h. Hose bibb on northwest corner
   i. Hose bibb on east side

3. a. _____
   b. _____
   c. _____
   d. _____
   e. _____
   f. _____
   g. _____
   h. _____
   i. _____

4. All supplies are to be valved. Where is this specified?

4. _____

5. What must plumbing contractor do regarding ductwork and electrical services?

5. _____
_____

6. What must the plumbing contractor install between connections of dissimilar metals?

6. _____

7. What is to be done to services to existing building?

7. _____
_____

8. What type water closet is specified?

8. _____
_____
_____

9. What is the number of the Dwyer Kitchen Unit specified, and what is to be replaced?

9. _____
_____
_____

10. Give the size of the building sewer.

10. _____

11. Where is this waste pipe to be connected?

11. _____
_____

12. What size waste pipe is used to drain the electric drinking fountain?

12. _____

13. Give the size of waste pipe serving the:
    a. Water closets
    b. Service sink
    c. Kitchen Dwyer Unit

13. a. _____
    b. _____
    c. _____

14. How many floor drains are there?

14. _____

15. Indicate the size of vent through roof for the:
    a. Kitchen Dwyer Unit
    b. Drinking fountain

15. a. _____
    b. _____

Unit 16
Air Conditioning System Blueprints

Air conditioning is the treatment of the air in a space (room or building) to make that space comfortable for those occupying the area. This treatment involves controlling the temperature, humidity (moisture in the air) and cleanliness of the air. In colder periods of the year, the air in the building is heated and moisture is added for comfort. In the warmer periods, the air is cooled and, often, moisture is removed for comfort.

To accomplish the desired air conditioning in a building, a heating and cooling system, and sometimes a system for cleaning and treating the air for desired humidity, are added. In this unit, we shall study heating and cooling systems and how these are shown on drawings.

Heating and Cooling Plans

The HEATING AND COOLING PLANS for residences and light commercial structures usually are prepared by the heating/cooling subcontractor. Generally, these plans are drawn on the floor plan of the structure for approval of the architect or homeowner. For larger commercial structures, the heating and cooling plans are prepared by an air conditioning engineer at the direction of the architect.

Symbols for heating and cooling systems are shown in Fig. 16-1. A symbol legend usually is included on the drawing showing the heating and cooling plan.

AIR CONDITIONING SYMBOLS

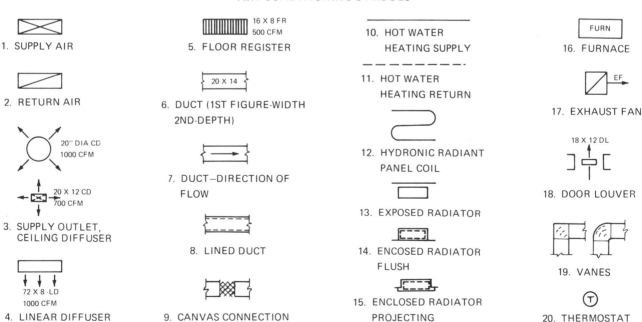

Fig. 16-1. Graphic symbols for air conditioning systems.

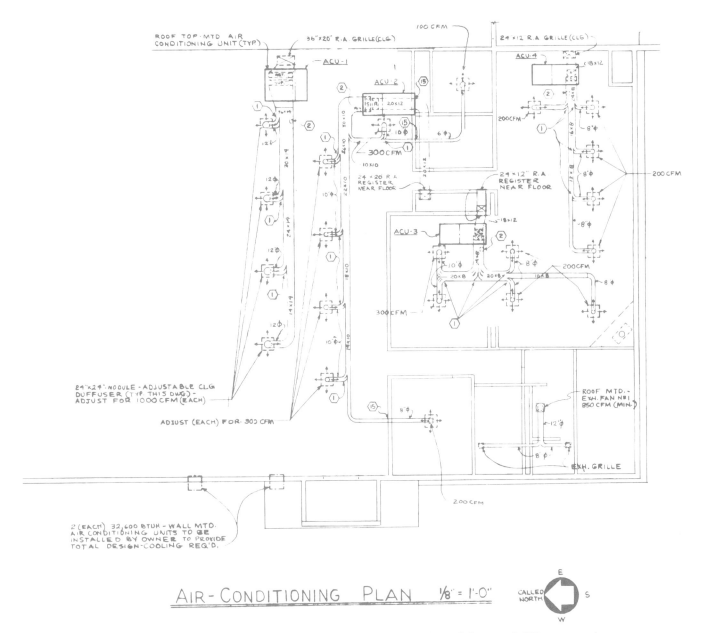

AIR-CONDITIONING PLAN 1/8" = 1'-0"

Fig. 16-2. Forced air heating and cooling plan drawing with sizes of ducts and diffusers noted.
(Henkel, Hovel & Schaefer Architects - Engineer)

Heating System

There are three basic types of heating systems used in new construction: forced air, hydronic (hot water) and electric radiant heating.

In a FORCED AIR SYSTEM, the heated air from the furnace or heat pump chamber is transferred by means of a motor-driven fan through a series of ducts (rectangular or round pipes) to registers or diffusers in the various rooms. See Fig. 16-2. Cool air is gathered through registers near the floor and returned to the heating unit through ducts and a filtering system to be reheated and recirculated.

In residences or light commercial structures, the forced air heating system is considered to be a closed circuit (no provision for outside air to be added) system. In large commercial structures, provision is made for the addition of "fresh" air from the outside and no air is returned from such rooms as kitchens, smoking rooms and rest rooms.

Sources of heat for forced-air systems are natural gas, liquid petroleum, gas, oil, coal or electricity such as radiant heating or the heat pump. Developments are also taking place in the use of solar energy as a fuel source in forced air systems.

In Fig. 16-2, note size of ducts specified. Some

154

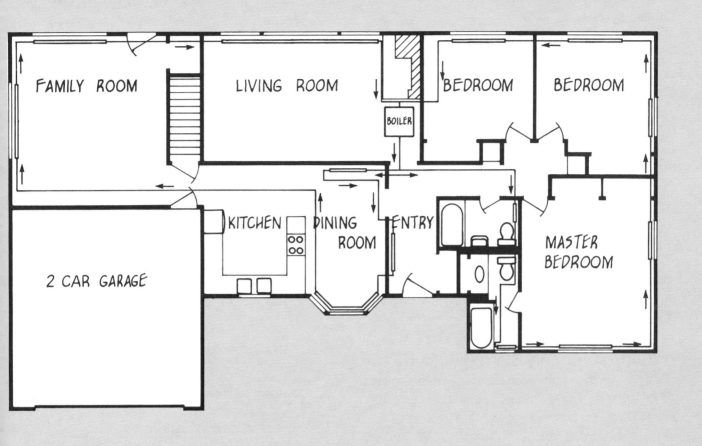

Fig. 16-3. A hydronic heating system layout, series-loop type.

commercial plans call out CFM (cubic feet per minute) to be delivered at a particular register.

HYDRONIC SYSTEM of heating involves the heating of water to a temperature of around 200 degrees F (92 C) in a boiler. Then, the hot water is circulated by a pump and pipe system to convectors in the spaces to be heated, Fig. 16-3. Hydronic systems for residences normally use the series-loop system to carry the heated water to the convectors, Fig. 16-3. For larger areas to be heated, the one-pipe system or two-pipe system are specified. The latter systems of piping for hydronic systems provide more uniform heat for larger areas.

When drawings are provided, they usually are superimposed over the floor plan or given in an isometric diagram.

ELECTRIC RADIANT HEATING for space heating usually is provided by wires embedded in the ceiling, walls, floor or baseboards. The wires are spaced on a grid pattern of approximately 1 1/2 inch centers and stapled to the gypsum lath of the ceiling or wall and covered with plaster. Floor cables are heavier. These are placed on a grid pattern 1 1/2 to 3 inches below the concrete surface.

Glass panel heaters are also available for ceiling and wall installation. These panels may be painted, papered or plastered.

Heating system drawings may be superimposed over the floor plans on a separate diagram provided with appropriate notes. When drawings are not provided, the amount of heat required for each space is noted on the floor plan.

Cooling Systems

Refrigeration type cooling systems may be classed as UNIT (window or wall mounted) coolers and REMOTE (refrigeration equipment located away from the area to be conditioned). Another type, called EVAPORATIVE (sometimes referred to as "desert" type) coolers were popular a number of years ago and gave way to the refrigeration type. Evaporative coolers are experiencing some renewed interest due to the energy shortage.

UNIT TYPE COOLERS, which are installed in a window or space provided in an exterior wall, are used to cool a room. Very little construction is involved in their installation. Units may be purchased to heat as well as cool the air.

REMOTE COOLING SYSTEMS have the condensing unit in a remote space away from the area to be cooled. The evaporator is in the main duct where the fan forces air past the cool coils and circulates the air to the rooms to be cooled. Another variation of this cooling system is with a remote condensing unit and evaporator, and a cooled brine or water is circulated to the heat exchangers in each room.

EVAPORATIVE COOLING SYSTEMS are most effective in dry climates where the relative humidity is low (preferably 20 percent or less). They will work, but less effectively, at higher levels of humidity. The evaporative system functions by moving air rapidly over a pad of excelsior that is kept moist by a water spray mist. After the cooled air passes through the pad, it is then carried through a duct system to the rooms. This system of cooling does raise the humidity in the space being cooled.

The floor plan drawing for an evaporative cooling system would be similar to the supply ducts of a forced air system. However, the evaporative system has larger ducts, and no return ducts are necessary since outside air is used.

Air Cleaners

Most heating and cooling systems provide a means of filtering the air that flows through the system. The filters usually have an adhesive or oil coat that collects the lint and dust particles. These filters are "throwaway" or washable types that can be re-coated and reused. Another effective air cleaner filter for heating and cooling systems is the electrostatic type shown in Fig. 16-4.

Most systems come equipped with a designated space for inserting the adhesive type filter. The electrostatic filter usually is a separate unit added to the system. It is noted on the heating and cooling plan, and detailed in the specifications.

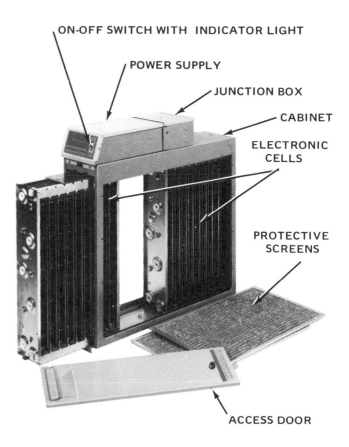

Fig. 16-4. An electrostatic air filter removes dust by electrically charging the particles. (Honeywell, Inc.)

Air Conditioning System Blueprints

Blueprint Reading Activity 16—1
RESIDENTIAL
AIR CONDITIONING SYSTEM BLUEPRINT

Refer to Print 16-1 in the Large Prints Folder to answer the following questions.

1. What is the scale of the air conditioning plan?   1. _____

2. What size and specifications are given for the A/C unit?   2. _____

3. Where is the A/C condensing unit located?   3. _____

4. What are the sizes of the refrigeration lines, and how are they brought into the house?   4. _____

5. Where is the hot air furnace located?   5. _____

6. What is the size of the main supply air duct coming off the furnace?   6. _____

7. Give the size of the supply air duct to the Den. How much air is to be supplied?   7. _____

8. Indicate how much air is to be supplied:
   a. To sewing room
   b. To dining room
   c. To kitchen
   d. To master bedroom

   8. a. _____
   b. _____
   c. _____
   d. _____

9. Give the number and sizes of the supply air registers:
   a. In living room
   b. In foyer
   c. In master bedroom
   d. In den

   9. a. _____
   b. _____
   c. _____
   d. _____

10. What size return air grille is specified, and where is it located?   10. _____

11. Give the specification for the exhaust fans.   11. _____

12. How many exhaust fans are shown on the plan, and where are they located?   12. _____

Refer to Print 16-2 in the Large Prints Folder to answer the following questions.

1. What is the scale for the A/C floor plan?

1. _____

2. What size is specified for the A/C unit?

2. _____
_____
_____

3. Above what room is the roof-mounted A/C unit located? (Refer to the floor plan for this building: Print 17-2a.)

3. _____

4. What is the size of the main return air duct adjacent to the A/C unit?

4. _____

5. What is the size of the fresh air register in this duct?

5. _____

6. What size are the supply air ducts that first enter below the roof?

6. _____
_____

7. Linear diffusers supplying air to the lobby:
   a. How many?
   b. Size?
   c. How much air?

7. a. _____
   b. _____
   c. _____

8. Diffusers supplying workroom:
   a. How many?
   b. Size?
   c. How much air?

8. a. _____
   b. _____
   c. _____

9. Exhaust fans:
   a. How many?
   b. Where located?
   c. Specifications?

9. a. _____
   b. _____
   c. _____

10. What size duct is used to supply the conference room-safe deposit booth area?

10. _____

11. How much air is to be supplied to the conference room?

11. _____

12. Where is the thermostat located?

12. _____

Unit 17
Electrical System Blueprints

The electrical blueprints for residential and light commercial buildings usually consist of the outlets and switches shown on the floor plan. Routing of the wires and number of circuits to be included generally are left to the electrical contractor and the controlling electrical codes for the area. For large industrial and commercial buildings involving elaborate electrical systems, detailed electrical diagrams are prepared by electrical engineers.

Electricians on the job must be able to read blueprints in order to plan where circuits will run in the basement or crawl spaces and attics. They also need to know the direction of run of floor and ceiling joists. To help you read and understand electrical blueprints, this unit will explain electrical terms, symbols, diagrams, codes and calculations for circuits.

Terms Used in Electrical Work

Following are the most common terms used in electrical construction:

AMPERE: The unit of measurement of electricity flowing through a conductor.

VOLTAGE: The electromotive force which causes current to flow through a conductor (wire).

WATT: The unit of measurement of electrical power. Amperes times volts equals watts. Most appliances are rated in watts.

CIRCUIT: Two or more conductors (wires) carrying electricity from the source (distribution panel) to an electrical device and return.

CIRCUIT BREAKER: A switching device that automatically opens a circuit when the circuit has been overloaded.

CONDUCTOR: A wire or material used to carry the flow of electricity.

CONDUIT: A channel or pipe, made of metal or other material, in which conductors are run. Required by code where conductors need protection.

CONVENIENCE RECEPTACLE: An outlet where current is taken from a circuit to serve electrical devices such as lamps, clocks, toasters, etc. Sometimes called receptacles.

SERVICE ENTRANCE: The conductors from the utility pole, service head and mast which bring the electrical current to the building through the meter to the distribution panel.

DISTRIBUTION PANEL: The insulated panel or box which receives the current from the source and distributes it through branch circuits to various points throughout the building. The panel contains the main disconnect switch and fuses or circuit breakers protecting each circuit.

GROUND WIRE: A wire connecting the circuit or device to the earth to minimize injuries from shock and possible damage from lightning.

Symbols Used in Electrical Drawings

Some of the more common symbols used on electrical drawings are shown in Fig. 17-1. Also, a schedule of symbols usually is given on one or more of the electrical plan sheets.

LIGHTING OUTLETS

◯ ─⊕	CEILING OUTLET
Ⓓ	DROP CORD
Ⓕ ─Ⓕ	FAN OUTLET
Ⓙ ─Ⓙ	JUNCTION BOX
Ⓛ PS ─Ⓛ PS	LAMP HOLDER WITH PULL SWITCH
Ⓧ ─Ⓧ	EXIT LIGHT OUTLET
Ⓛ ─Ⓛ	OUTLET CONTROLLED BY LOW VOLTAGE SWITCHING WHEN RELAY IS INSTALLED IN OUTLET BOX
▢	SURFACE OR PENDANT INDIVIDUAL FLUORESCENT FIXTURE
▢OR	RECESSED INDIVIDUAL FLUORESCENT FIXTURE
Ⓡ ─Ⓡ	RECESS INCANDESCENT

SIGNALING SYSTEM OUTLETS
RESIDENTIAL OCCUPANCIES

▢•	PUSH BUTTON
▱	BUZZER
◖	BELL
◀	OUTSIDE TELEPHONE
◁	INTERCONNECTING TELEPHONE
▢D	ELECTRIC DOOR OPENER
▢CH	CHIME
▢TV	TELEVISION OUTLET
Ⓣ	THERMOSTAT

RECEPTACLE OUTLETS

─⊖	DUPLEX RECEPTACLE OUTLET
─⊖WP	WEATHERPROOF RECEPTACLE OUTLET
─⊕	TRIPLEX RECEPTACLE OUTLET
─⊕	QUADRUPLEX RECEPTACLE OUTLET
─⊜	DUPLEX RECEPTACLE OUTLET–SPLIT WIRED
─△	SINGLE SPECIAL–PURPOSE RECEPTACLE OUTLET
─⊖R	RANGE OUTLET
─▲DW	SPECIAL PURPOSE CONNECTION
Ⓒ	CLOCK HANGER RECEPTACLE
⊙	FLOOR SINGLE RECEPTACLE OUTLET
⊞	UNDERFLOOR DUCT AND JUNCTION BOX FOR TRIPLE, DOUBLE OR SINGLE DUCT SYSTEM AS INDICATED BY NUMBER OF PARALLEL LINES

PANELS CIRCUITS AND MISCELLANEOUS

▬	LIGHTING PANEL
▨	POWER PANEL
────	WIRING, CONCEALED IN CEILING OR WALL
─ ─ ─ ─	WIRING, CONCEALED IN FLOOR
◀───	CONDUIT RUN TO PANEL BOARD
─╫──	*INDICATES NUMBER OF CONDUCTORS
▢┐	EXTERNALLY OPERATED DISCONNECT SWITCH

SWITCH OUTLETS

S	SINGLE POLE SWITCH
S 2	DOUBLE POLE SWITCH
S 3	THREE WAY SWITCH
S 4	FOUR WAY SWITCH
S K	KEY OPERATED SWITCH
S P	SWITCH AND PILOT LAMP
S WCB	WEATHERPROOF CIRCUIT BREAKER
S WP	WEATHERPROOF SWITCH
S L	SWITCH FOR LOW VOLTAGE SWITCHING SYSTEM
S T	TIME SWITCH
Ⓢ	CEILING PULL SWITCH
─⊖s	SWITCH AND SINGLE RECEPTACLE
─⊖s	SWITCH AND DOUBLE RECEPTACLE
S CB	CIRCUIT BREAKER
S RC	REMOTE CONTROL SWITCH
S F	FUSED SWITCH
S LM	MASTER SWITCH FOR LOW VOLTAGE SWITCHING SYSTEM
S D	AUTOMATIC DOOR SWITCH

*Indicates number of conductors (in this case, 4). Any circuit without cross hatches indicates two-conductor circuit. Some electrical engineers show number of hot conductors with full marks; neutral conductors with half marks (─╫╵─ = 3 hot conductors, 1 neutral).

Fig. 17-1. Common symbols used on electrical plan drawings and diagrams.

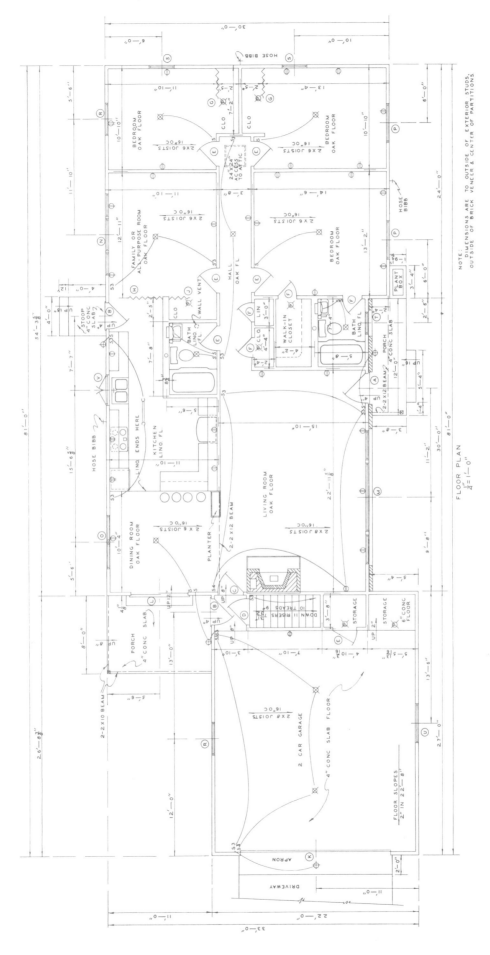

Fig. 17-2. An electrical plan drawing for a residence.

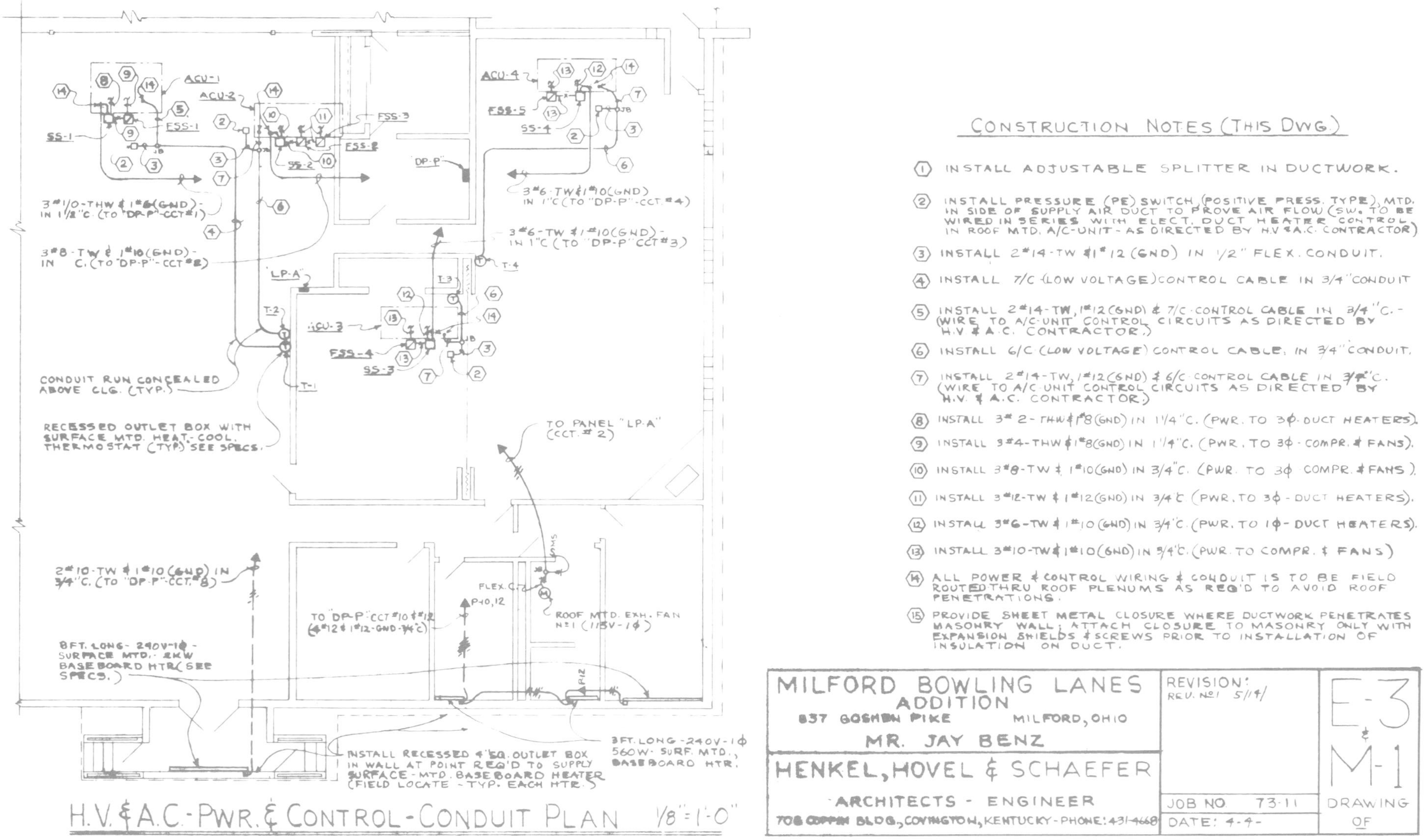

CONSTRUCTION NOTES (THIS DWG.)

1. INSTALL ADJUSTABLE SPLITTER IN DUCTWORK.

2. INSTALL PRESSURE (PE) SWITCH (POSITIVE PRESS. TYPE), MTD. IN SIDE OF SUPPLY AIR DUCT TO PROVE AIR FLOW (SW. TO BE WIRED IN SERIES WITH ELECT. DUCT HEATER CONTROL) IN ROOF MTD. A/C UNIT. AS DIRECTED BY HV & A/C CONTRACTOR)

3. INSTALL 2 #14-TW & 1 #12 (GND) IN 1/2" FLEX. CONDUIT.

4. INSTALL 7/C (LOW VOLTAGE) CONTROL CABLE IN 3/4" CONDUIT.

5. INSTALL 2 #14-TW, 1 #12 (GND) & 7/C CONTROL CABLE IN 3/4" C. (WIRE TO A/C UNIT CONTROL CIRCUITS AS DIRECTED BY HV & A.C. CONTRACTOR).

6. INSTALL 6/C (LOW VOLTAGE) CONTROL CABLE IN 3/4" CONDUIT.

7. INSTALL 2 #14-TW, 1 #12 (GND) & 6/C CONTROL CABLE IN 3/4" C. (WIRE TO A/C UNIT CONTROL CIRCUITS AS DIRECTED BY HV & A.C. CONTRACTOR).

8. INSTALL 3 #2 - THW & 1 #8 (GND) IN 1 1/4" C. (PWR. TO 3∅ DUCT HEATERS).

9. INSTALL 3 #4 - THW & 1 #8 (GND) IN 1 1/4" C. (PWR. TO 3∅ COMPR. & FANS).

10. INSTALL 3 #8-TW & 1 #10 (GND) IN 3/4" C. (PWR. TO 3∅ COMPR. & FANS).

11. INSTALL 3 #12-TW & 1 #12 (GND) IN 3/4 C. (PWR. TO 3∅ DUCT HEATERS).

12. INSTALL 3 #6-TW & 1 #10 (GND) IN 3/4" C. (PWR. TO 1∅ DUCT HEATERS).

13. INSTALL 3 #10-TW & 1 #10 (GND) IN 3/4 C. (PWR. TO COMPR. & FANS)

14. ALL POWER & CONTROL WIRING & CONDUIT IS TO BE FIELD ROUTED THRU ROOF PLENUMS AS REQ'D TO AVOID ROOF PENETRATIONS.

15. PROVIDE SHEET METAL CLOSURE WHERE DUCTWORK PENETRATES MASONRY WALL & ATTACH CLOSURE TO MASONRY ONLY WITH EXPANSION SHIELDS & SCREWS PRIOR TO INSTALLATION OF INSULATION ON DUCT.

E-3		
M-1		
DRAWING		
		OF

MILFORD BOWLING LANES	REVISION:
ADDITION	REV. N.º 1 5/14/
837 GOSHEN PIKE MILFORD, OHIO	
MR. JAY BENZ	
HENKEL, HOVEL & SCHAEFER	
- ARCHITECTS - ENGINEER	JOB NO. 7311
708 COPPIN BLDG., COVINGTON, KENTUCKY - PHONE: 431-4668	DATE: 4-4-

H.V. & A.C.-PWR. & CONTROL-CONDUIT PLAN   1/8" = 1'-0"

Fig. 17-3. A wiring diagram   (Henkel, Hovel & Schaefer, Architects - Engineer)

Any standard symbol may be used to designate some variation of standard equipment by the addition of lower case subscript lettering to the symbol. This would be identified in the schedule of symbols and, if necessary, further described in the specifications.

Electrical Plan Drawings

The electrical layout showing the distribution panel, convenience outlets, switches, lights, etc., usually is placed on a copy of the floor plan and labeled "Electrical Plan." See Fig. 17-2. Broken lines indicate which outlets and switches are connected, however, the path of the wire is not necessarily where the lines are drawn.

For larger construction jobs, an electrical plan may be prepared for outlets, another for lighting and still another for the service entrance. These plans, together with the set of specifications, detail the electrical work to be done and the materials and fixtures to be used.

A WIRING DIAGRAM, Fig. 17-3, is more detailed than an electrical plan drawing. It shows the outlets and switches, and also shows how the circuits in the building are arranged.

A CIRCUIT is the path of electricity from a source (distribution panel) through the components (electrical outlets, lights) and back to the source. Circuits are numbered on the diagram and connected by a heavy line, ending in an arrow that indicates the circuit is connected to the distribution panel at this point, Fig. 17-3.

Electrical Circuits

Electricity is brought into the building by way of the service entrance through the meter and on to the distribution panel. For most residences and light commercial buildings, one distribution panel is sufficient. Large commercial buildings make use of feeder circuits to further distribute the electricity to sub-distribution panels. Larger conductor sizes should be used in feeder circuits. This helps to avoid excessive voltage drop in branch circuits which otherwise might be in excess of 75 to 100 feet in length.

BRANCH CIRCUITS may be classified as:
1. General lighting circuits used primarily for lighting and small portable appliances such as radios, clocks, TV sets and vacuum cleaners.
2. General appliance circuits used for those outlets along the kitchen counter serving toasters, waffle irons, mixers and other appliances. These circuits are also used for home workshops.
3. Individual appliance circuits used for major appliances which require large amounts of electricity, such as range-oven, washer, dryer and water heater.

CALCULATIONS FOR CIRCUITS should be made by the electrical contractor if the number of circuits has not been indicated on the drawings or in the specifications. Circuits serving lights and small appliances usually are planned to serve 2400 watts (20 amps x 120 volts = 2400 watts). These circuits normally are wired with # 12 gauge wire.

Circuits serving individual appliances would be wired with a size wire and circuit breaker or fuse to safely carry the amps required by the appliance. For example, an individual circuit for a 240 volt range-oven using 12,000 watts would require at least a 50 amp circuit ($\frac{12,000 \text{ watts}}{240 \text{ volts}}$ = 50 amps). This circuit would be sized at 60 amps, using a # 6 gauge wire.

Remote Control Systems

The use of a low voltage (24 volts) wiring system makes it possible to control the switching of any light or outlet in a building from any location in the building. The lights or outlets are wired using standard wire sizes of # 12 or # 14. The switches are operated through a low voltage system using bell wire. The switch (located anywhere in the building) in the low voltage system activates a relay (electrically operated switch) at the outlet, which turns the device ON or OFF. This system is gaining popularity because of its flexibility.

Codes

It is the architect's responsibility, with the assistance of his consulting engineers, to design a building to meet existing codes. This includes the provisions of the National Electrical Code and any state or local codes. In the case of residential or light commercial buildings where no detailed electrical diagram or specifications are provided, the electrical subcontractor or individual is responsible for meeting the provisions of the codes.

Types and Sizes of Conductors

Conductors (wires) in most building projects are specified as copper, although aluminum is also used. Sizes of wire are designated by gauge numbers based on the diameter of the wire. Some wire sizes are shown in Fig. 17-4. Note that as the size number decreases, the wire size increases. Most residential wiring calls for # 12 size wire in copper or # 10 in aluminum. Number 14 copper is the smallest conductor permitted in branch circuits by the National Electrical Code.

The National Electric Code uses letters to designate the type of conductor insulation. This governs the use of electrical conductors. Some of the more common designations are listed in Fig. 17-5. For more specific applications, check the latest edition of the National Electric Code. The letters AWG and MCM are found on some electrical blueprints. AWG refers to American Wire Gauge, which is a means of specifying wire diameter. MCM refers to Thousand Circular Mills, which designates the thickness of the insulation.

Conduit

CONDUIT is classified as rigid (pipe) or electrical metal tubing (EMT). EMT also is referred to as thin-wall.

TYPES OF CONDUCTOR INSULATIONS

LETTER DESIGNATION	TYPE OF INSULATION
RH	Heat-Resistant Rubber
RHH	Heat-Resistant Rubber
RHW	Moisture and Heat Resistant Rubber
RUH	Heat Resistant Latex Rubber
RUW	Moisture Resistant Latex Rubber
T	Flame Retardant Thermoplastic
TW	Flame Retardant, Moisture Resistant Thermoplastic
THHN	Flame Retardant, Heat Resistant Thermoplastic
THW	Flame Retardant, Moisture and Heat Resistant Thermoplastic
THWN	Flame Retardant, Moisture and Heat Resistant Thermoplastic
XHHW	Moisture and Heat Resistant Cross-Linked Synthetic Polymer
MTW	Moisture, Heat and Oil-Resistant Thermoplastic
TFE	Extruded Polytetrafluorethylene
TA	Thermoplastic and Asbestos
MI	Mineral Insulation (Metal Sheathed) Magnesium Oxide

Fig. 17-5. Types of wire and cable insulations.
(National Electric Code)

Most rigid conduit is galvanized or enameled steel with threaded joints or threadless compression fittings. The conduit comes in 10 foot lengths and may be given a radius bend with the proper equipment. Rigid conduit inside dimensions are nominal sizes (1 inch is actually 1.049 inches). Rigid conduit also is available in nonmetallic materials of asbestos cement, polyvinyl chloride (PVC) and polyethylene.

Electrical metal tubing (thin-wall) has thinner walls than rigid conduit and is more easily bent to shape. It usually is made of galvanized steel, although it is available in bronze for use in corrosive atmospheres. Like rigid metallic conduit, thin-wall conduit comes in nominal size inside dimensions. It is available in sizes up to 4 inches.

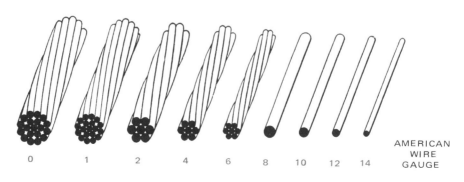

AMERICAN WIRE GAUGE

0   1   2   4   6   8   10   12   14

Fig. 17-4. Relative sizes of electrical conductors (wires).

USE AND/OR AREA SERVED	C/B	CIR. NO	LOAD ØA	ØB	CIR. NO	C/B	USE AND/OR AREA SERVED
GENERAL - DEN	20/1	1			2	20/1	GENERAL - B.R., BATH
" LIVING RM		3			4		" SEWING RM
LITES - FRONT		5			6		" MSTR B.R.
RECEPS - PLANTERS		7			8		LITES - UTIL, KITCHEN
LITES - GARAGE, 2RM		9			10		FURNACE FAN
RECEPS - GARAGE		11			12		WATER SOFTENER CONT.
GARAGE DOOR		13			14		KITCHEN RECEPS
SPARE		15			16		" "
WASHER		17			18		DISPOSEL, D.W
FREEZER		19			20		POOL LITE
FUTURE SITE LITES	20/1	21			22	20/1	POOL PUMP
CLOTHES DRYER	30/	23			24	50/	RANGE
	/2	25			26	/2	
WATER HEATER	30/	27			28	40/	OVEN
	/2	29			30	/2	
RELAY PANEL POWER	20/1	31			32	20/1	SPARE
GARAGE DOOR MOTOR		33			34		SPARE
SPARE		35			36		SPARE
SPARE	20/1	37			38	20/1	SPARE
SPACE ONLY		39			40		SPACE ONLY
" "		41			42		" "

PANEL A — TYPE NQO — MOUNTING RECESS — 120/240V 1Ø 3W — MAINS 225A

Fig. 17-6. Typical branch circuit schedule.   (Robert Helgeson, Architect)

FIXTURE SCHEDULE					
MARK	MANUFACTURE	CAT. Nº	WATTAGE	FINISH	REMARKS
A	LITHONIA	2G 440-A12	4-40	WHITE	GRID CEILING ACRYLIC.
B	PRESCOLITE	4020	100	BLACK	
C	PRESCOLITE	4452	4-60	BLACK	
D	PRESCOLITE	7HV	75-2-30	WHITE	
E	PRESCOLITE	1252-916	75-R-30	BRONZOTIC	
F	PRESCOLITE	540	100W	WHITE	8" TUBE
G		RECESSED CEILING MT.	100W		SUITABLE FOR USE IN SAUNA
H	LITHONIA	G 240-A12	2-40	WHITE	SURFACE
J.	SPAULDING	832	60	BRONZE ANOD.	WALL BRACKET
K	PRESCOLITE	37D-B	2-60	ALUM	6" A.F.F.
L.	PRESCOLITE	93020	175W-MV	BLACK	
N.	PRESCOLITE	4220	100	BLACK	
O	ALLOWANCE	OF $250.00/OUTLET			OUTLET ONLY
P.	————	————	100		PORCELAIN SOCKET
R	PRESCOLITE	1171-900	2-50 R20	BRONZOTIC	

Fig. 17-7. A lighting fixture schedule for an office building.   (John J. Ross, Architect)

Conductors also may be run through metal or plastic surface raceways (channels) with snap-on covers. Raceways are also made to be placed in concrete floors, which have surface covered junction boxes at regular intervals so that electrical connections may be made where needed.

Schedules

SCHEDULES usually are furnished for the branch circuits and for the lighting fixtures. The branch circuit schedule lists the individual circuits by number, location and electrical devices included in the circuit. See Fig. 17-6. Some schedules also show the ampere rating of the circuit breaker switch and the size conductor to be used in the branch circuit.

The lighting fixtures for many residences are purchased and furnished by the owner. In some cases, an allowance of a certain dollar amount is included in the contract. The owner selects fixtures up to this amount and purchases those beyond the allowance. In other residences and commercial buildings, the lighting fixtures are selected by the architect and are identified as to type, make, mounting instructions and any special features to be included. Fig. 17-7 shows a fixture schedule.

Electrical System Blueprints

Refer to Print 17-1 in the Large Prints Folder to answer the following questions.

1. What type of wiring is to be used on the job?

1. _____
   _____

2. What type switches and receptacles are to be supplied?

2. _____
   _____

3. There are four special purpose convenience connectors in the kitchen. What do they serve?

3. _____

4. What type switches control the lights at the end of the dining room and where are they located?

4. _____
   _____

5. How are the ceiling lights in the dining room controlled? What type are they?

5. _____
   _____
   _____

6. How many wall mounted surface fixtures are there? What type are they?

6. _____
   _____

7. What is the meaning of the number alongside a lighting or convenience outlet, such as the number "32" next to the dining room ceiling lights?

7. _____
   _____

8. Explain what is required at A in the living room. What circuit are they on?

8. _____
   _____

9. Are there other devices on circuit #24? If so, where?

9. _____
   _____

10. What type fixture is at B?

10. _____
    _____

11. The light above the stair landing is of what type and how is it controlled?

11. _____
    _____
    _____

12. What type light is at C and where are they controlled? What circuit are they on?

12. _____
    _____
    _____

13. Who is to supply the light fixtures? Who is to install?

13. _____
    _____

Blueprint Reading Activity 17−2
ELECTRICAL SYSTEM BLUEPRINTS
OFFICE BUILDING

Refer to Prints 17-2a and 17-2b in the Large Prints Folder to answer the following questions.

Panel A - Receptacles

1. What circuits serve the following:
   a. Water heater?
   b. Dwyer Kitchen Unit?
   c. Receptacles in safe deposit booths?
   d. Electric drinking fountain?
   e. Weatherproof receptacle on column on south?

   1. a. _____
      b. _____
      c. _____
      d. _____
      e. _____

2. What is 8 in the lounge and what circuit controls it?

   2. _____

3. What circuit serves the accounting machines on the teller line?

   3. _____

4. Power is brought to the vault by which circuit?

   4. _____

5. Interpret information given at A.

   5. _____
      _____

6. Interpret information given at B.

   6. _____
      _____
      _____

7. Interpret information given at C.

   7. _____
      _____
      _____

Panel A - Lighting Plan

8. What type of light fixtures are called for at D and on what circuit are they?

   8. _____
      _____

9. What is to be done about fixture at E?

   9. _____
      _____

10. What lights are to be provided for Drive-In-Teller Canopy, and where is the control switch?

    10. _____
        _____
        _____
        _____

11. Interpret information given at F.

11. _____
_____
_____

12. Lights on column on South walk:
    a. Type?
    b. Where positioned?
    c. What circuit?
    d. How controlled?

12. a. _____
    b. _____
    c. _____
    d. _____

13. Identify device at G.

13. _____
_____

14. Where is the conduit at H run?

14. _____

15. What is the fixture at J and where is it positioned?

15. _____
_____

Unit 18
Unit Masonry Construction Blueprints

Unit masonry construction, in contrast to poured concrete (monolithic), is made up of a number of smaller units held together with a bonding material known as mortar. In this unit of the write-in text, you will study types of unit masonry construction and learn how each type lends itself to simplified construction procedures and a variety of design textures.

Various walls constructed of unit masonry are shown in Fig. 18-1:
a. SOLID WALLS with same material used through entire wall thickness.
b. SINGLE WYTHE (continuous vertical section one unit thick) of brick or stone, backed up with concrete block or tile.
c. CAVITY WALLS, which have a space between the outer and inner courses of material.
d. VENEERED WALLS of brick or stone, backed up with wood frame.

Types of Unit Masonry

Unit masonry includes the following common types: brick, concrete block, stone and structural clay tile.

Brick Masonry

Brick masonry walls may be laid up as SOLID BRICK WALLS of 8 inches or more in thickness. See (a) in Fig. 18-1. The inner wythe of brick usually is a building (common) brick, and the two courses are tied together in a bond pattern (discussed later in this unit).

SINGLE WYTHE BRICK WALLS are common in commercial construction with the face brick backed up by less expensive concrete block, hollow brick or structural clay tile to complete the wall thickness. See (b) in Fig. 18-1.

Brick walls are constructed as CAVITY WALLS when the inside surface of the wall is to be of exposed brick. The cavity in the wall, shown at (c) in Fig. 18-1, lessens the transfer of moisture, temperature and sound. Note how the two wythes are tied together with reinforcing wire.

In addition, the surface of the rear brick wall next to the cavity would be parged (coated with mortar) to provide additional waterproofing. A metal flashing also is installed to collect moisture and drain it through "weep" holes in the mortar to the outside.

BRICK VENEERED WALLS of one wythe are also laid as a VENEER of "skin" around a frame building. Note at (d) in Fig. 18-1 how the foundation wall is recessed to provide for the brick to extend lower on the foundation.

SYMBOLS FOR BRICK: Brick is indicated on plan and section drawings with 45 degree cross-hatch lines. For common brick, the lines are widely spaced; for face brick, the spacing is narrower. See Fig. 18-2. Fire brick is shown on plan drawings with

Unit Masonry Construction Blueprints

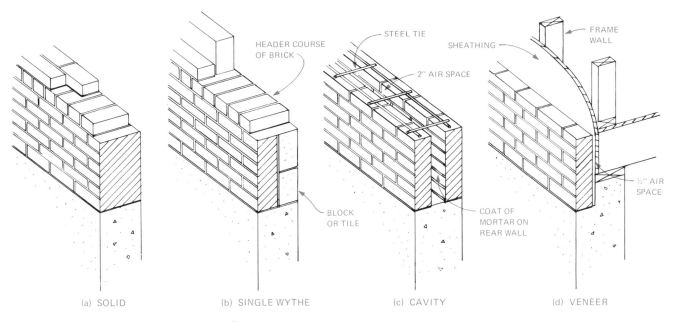

Fig. 18-1. Types of unit masonry walls.

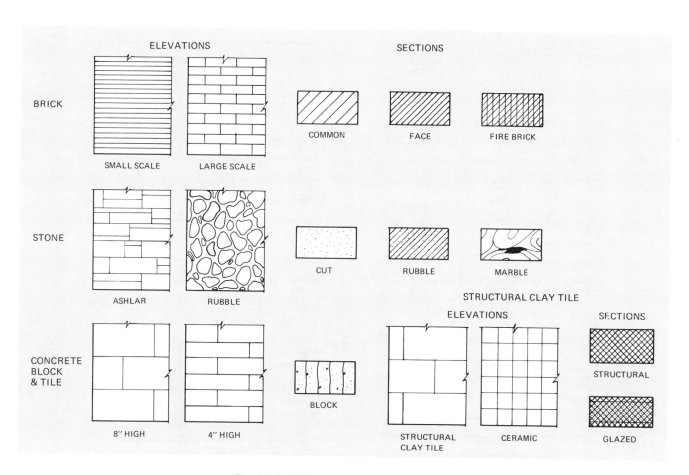

Fig. 18-2. Unit masonry material symbols.

the usual 45 degree lines indicating brick, plus vertical lines which designate it as fire brick.

On elevation drawings, brick normally is shown by horizontal lines and a notation on the type of brick such as "face" or "used brick." Some architects, to save time, will draw horizontal lines to indicate brick around the outer surface only of the brick walls.

TYPES OF BRICK BONDS: There are several types of brick bonds, so construction workers should be familiar with those more widely used. A BOND is the pattern of brick that repeats itself when laid in courses (rows). Bonds are designed for appearance or to add strength to a structural wall or to tie a wythe wall (a veneer section of wall one brick thick) to a backup wall.

The bonds most widely used are common bond, running bond, English bond, English cross bond, Flemish bond and stacked bond. See Fig. 18-3. When bricks are laid end to end in a course, they are called STRETCHERS as in the running bond. In

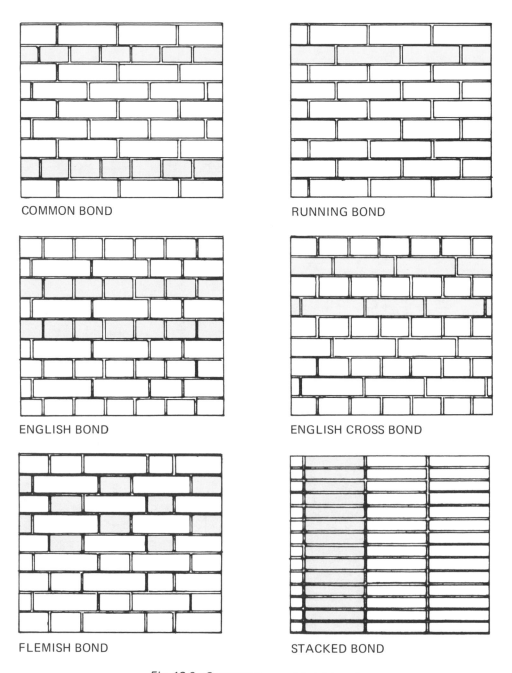

COMMON BOND

RUNNING BOND

ENGLISH BOND

ENGLISH CROSS BOND

FLEMISH BOND

STACKED BOND

Fig. 18-3. Common types of brick bonds.

some bonds, such as the common bond (sometimes called American bond), every sixth or seventh course is turned 90 degrees for appearance or to tie a structural wall together. Bricks laid in this manner are called a HEADER COURSE. There are other bond patterns for special effects such as HERRING BONE and BASKETWEAVE that may be called for on drawings.

BRICK POSITIONS: You may have observed the various positions that brick have been laid in a wall. These positions are planned by the architect to develop a certain design or style in the building as well as to add to the structural strength of the brickwork. Each position has a name that identifies it in a wall. For example, the most common position is the "stretcher," Fig. 18-4, which is laid in a flat position, lengthwise with the wall. Bricks in stretcher positions make up the larger portion of most walls.

The "header" is laid flat across the wall thickness to tie the wall together as well as to add interest to a pattern. Brickmasons, as well as other craftsworkers, must know these positions in order to construct the building according to plan.

CONCRETE BLOCK MASONRY comes in many varieties including the plain, colored surfaces, special design features and the popular "slump" block that simulates adobe brick or stone. See Fig. 18-5. Concrete block, like brick, may be laid in a variety of bond patterns to produce strength as well as a design effect.

SYMBOLS for concrete block in plan and section views are the same as concrete with the addition of lines crosswise of the run, Fig. 18-2. The elevation view symbol for concrete block is the same as for poured concrete with lines added to represent the block pattern.

Stone Masonry

STONE MASONRY may be laid in solid walls of stone or in single wythe backed with concrete block or tile, or it may be laid as veneer. Walls of stone masonry are classified according to shape and surface finish of the stone such as rubble, ashlar and cut stone.

RUBBLE masonry consists of stones as they come from the quarry or are gathered from a field

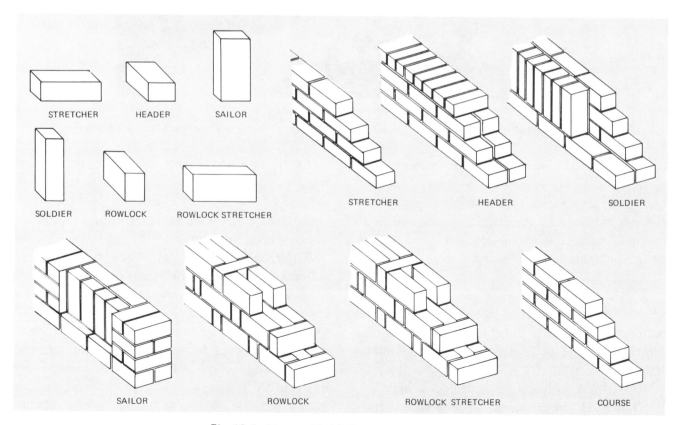

Fig. 18-4. Names of brick "positions" in a wall.

Fig. 18-5. This type of concrete block is called "slump" block. (National Concrete Masonry Association)

or stream. Such stones may be smooth with rounded edges or rough and angular. Fig. 18-6 illustrates patterns of rubble stone masonry. The random rubble wall consists of stones laid in an irregular pattern with varying sizes and shapes. Other rubble patterns are coursed, mosaic and strip.

ASHLAR masonry walls are squared stones which have been laid to a random or uniform pattern but not cut to dimensions as specified on a shop drawing, Fig. 18-6. The patterns of ashlar walls are:

REGULAR, with uniform continuous height.

STACKED, which tends to be a columnar in form.

BROKEN RANGE with squared stones of different sizes laid in uniform courses, but broken range within a course.

RANDOM RANGE, in which course and range do not remain uniform.

RANDOM ASHLAR, in which course is not uniform and ends are broken, not square.

CUT STONES, also known as dimensional stones, are cut and finished at the mill to meet the specifications of a particular construction job. Each stone is numbered for location on the job per the shop drawings furnished. Unlike ashlar masonry, which is laid largely at the design of the mason, cut stones are laid according to the design of the architect.

Bond Beams

Concrete masonry walls are usually reinforced horizontally and vertically at points of stress by constructing a BOND BEAM within the wall. This is done by pouring a watery concrete mixture called "grout" around reinforcing steel inserted in the units. Special channel blocks are used to form the horizontal bond beams, using reinforcing steel and mortar or grout. Vertical bond beams are formed by inserting reinforcing bars in a vertical cell after the wall is laid, then filling the cell with grout.

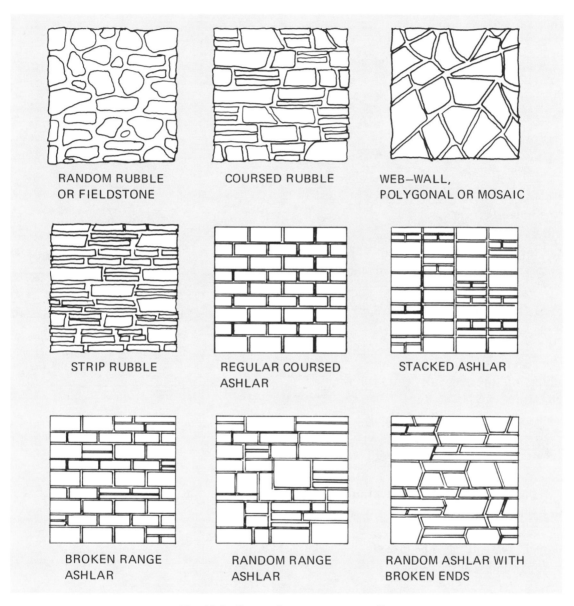

Fig. 18-6. Types of stone masonry walls.

175

Refer to Prints 18-1a through 18-1c in the Large Prints Folder to answer the following questions.

1. What is the scale of:
   a. The floor plan?   1. a. _____
   b. The transverse section?   b. _____

2. Give the dimensions for the following:
   a. Foundation exterior length?   2. a. _____
   b. Foundation exterior width?   b. _____
   c. Exterior length from stud face to stud face?   c. _____
                                                    _____
   d. Exterior width from stud face to stud face?   d. _____

3. Give the distance from the stud face at corner
   A of the house to:
   a. Center of garage door opening.   3. a. _____
   b. Stud face at B?   b. _____
   c. Center of living room window opening?   c. _____
   d. Center of front door opening.   d. _____
   e. Center of opening for garage side door?   e. _____

4. Give the distance from the stud face at corner
   C to:
   a. Center of opening for rear window of   4. a. _____
      bedroom?
   b. Family room door center opening.   b. _____

5. What material is used for the exterior wall?   5. _____

6. What type masonry unit construction is the   6. _____
   exterior wall?

7. How much air space is to be provided between   7. _____
   the masonry and wall sheathing?

8. How far below the top of the foundation wall   8. _____
   is the recess for the stone veneer?

9. From the "zero" (EL 100'-0") elevation as-   9. _____
   signed to the base of the stone veneer, how   _____
   high does the stone extend?   _____
                                                    _____
                                                    _____
                                                    _____

10. Is the construction of the garage exterior wall different from the house exterior walls?

10. _____

11. Although no design pattern is specified for the laying of the cut stone, what pattern is suggested in the elevation drawing?

11. _____

12. Does the cut stone continue over the door and window openings?

12. _____

13. What is the overall plan size of the fireplace and barbecue (excluding the hearth)?

13. _____

14. What are the dimensions of the hearth?

14. _____

15. What material is used for the fireplace exterior?

15. _____

16. What material is used to line the combustion chamber?

16. _____

17. What material is used as backup for the stone and fire brick?

17. _____

18. Give the size of the lintel used across the opening of:
    a. The fireplace?
    b. The barbecue?

18. a. _____
    b. _____

19. What are the dimensions of the fireplace opening:
    a. Height?
    b. Width?
    c. Depth?

19. a. _____
    b. _____
    c. _____

20. What is the size of the flue liner?

20. _____

Blueprint Reading Activity 18—2
UNIT MASONRY CONSTRUCTION
OFFICE BUILDING

Refer to Prints 18-2a through 18-2e in the Large Prints Folder to answer the following questions.

1. What is the scale of the elevation drawings?

2. Explain the construction of the wall at A on Sheet 5.

3. How is wall B constructed?

4. Explain the construction at wall C.

5. Describe the construction of the wall at D.

6. Describe the wall section E at the drive-thru drawer.

7. How are the 2'-8" square columns at F and G constructed?

7. _____

8. There are openings in the masonry wall above the north wall of the vault. Describe these openings and their location.

8. _____

9. Describe the location and support for the roof beam extending between the two Pylons at the Drive Thru.

9. _____

10. What are the construction details of the wall at H?

10. _____

11. Describe the construction of the wall at J.

11. _____

12. Describe the construction of the wall at K.

12. _____

Unit 19
Welding Construction Blueprints

Welding has become one of the construction industries' principal means of fastening members in structural steel and reinforced concrete work, Fig. 19-1. The American Welding Society has developed and adopted standard procedures for using symbols to indicate the exact location, size, strength, geometry and other information necessary to describe the weld required. The welding symbols studied in this Unit will assist you in reading and interpreting drawings involving welding processes.

Elements of Welding Symbols

It is important to distinguish between the WELD SYMBOL and the WELDING SYMBOL. The weld symbol indicates the type of weld only. The welding symbol consists of the following elements:

The REFERENCE LINE is the horizontal line (may appear vertically on some prints) portion of a welding symbol, Fig. 19-2, which is joined by an arrow and a tail.

An ARROW is used to connect the welding symbol reference line to one side of the joint. See Fig. 19-2. This is considered the ARROW SIDE of the joint. The side opposite the arrow is termed the OTHER SIDE of the joint.

The TAIL SECTION shown in Figs. 19-2 and 19-3 is used for designating the welding specification process (for abbreviations, see page 326) or other reference such as an industry specification.

Fig. 19-1. A construction welder at work.

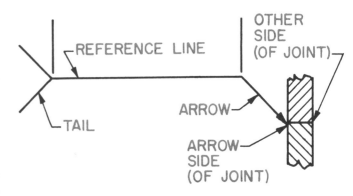

Fig. 19-2. Basic welding symbol.

Fig. 19-3. Tail section designating arc welding.

BASIC WELD SYMBOLS for various types of welds are shown in Fig. 19-4. A more comprehensive list of weld symbols and their applications are shown in page 240.

LOCATION OF WELDS are shown by their placement on the reference line. Weld symbols placed on the side of the reference line nearest the reader refer to welds on the ARROW SIDE of the joint, Fig. 19-5. Weld symbols on the reference line side away from the reader indicate welds on the OTHER SIDE of the joint. Weld symbols on both sides of the reference line indicate welds on BOTH SIDES of the joint.

DIMENSIONS OF WELDS are drawn on the same side of the reference line as the weld symbol, Fig. 19-6(a). When the dimensions are covered by a general note, such as "ALL FILLET WELDS 3/8" IN SIZE UNLESS OTHERWISE NOTED," the welding symbol need not be dimensioned, Fig. 19-6(b). When both welds have the same dimensions, one or both may be dimensioned, Fig. 19-6(c). The pitch of staggered intermittent welds is shown to the right of the length of the weld, Fig. 19-6(d).

SUPPLEMENTARY AND OTHER SYMBOLS used with the welding symbol to further specify the type of weld are discussed under the following:

CONTOUR SYMBOL shown next to the weld symbol indicates fillet welds that are to be flat-faced, Fig. 19-7(a); convex, Fig. 19-7(b); or con-cave-faced, Fig. 19-7(c).

GROOVE ANGLE, Fig. 19-8(a), is shown on the same side of the reference line as the weld symbol. The size (depth) of groove welds is shown to the left of the weld symbol, see Fig. 19-8(b). The root opening of groove welds, Fig. 19-8(c), is shown inside the weld symbol.

SPOT WELDS are called out for size in diameter, Fig. 19-9(a), strength in pounds, Fig. 19-9(b); pitch (center to center), Fig. 19-9(c); and number of spot welds, Fig. 19-9(d).

WELD-ALL-AROUND symbol, Fig. 19-10, indicates that the welds extend completely around a joint.

ARROW SIDE   OTHER SIDE   BOTH SIDES

Fig. 19-5. Location of welds.

Groove							
Square	‖ Scarf*	V	Bevel	U	J	Flare-V	Flare-bevel

Fillet	Plug or slot	Spot or projection	Seam	Back or backing	Surfacing	Flange	
						Edge	Corner

*Used for brazed joints only

Fig. 19-4. Basic weld symbols.   (American Welding Society)

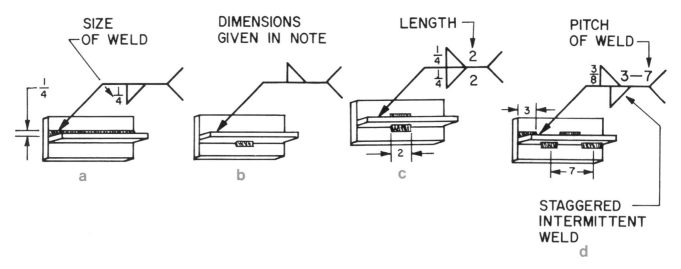

Fig. 19-6. Examples of accepted methods of depicting dimensions of welds. Dimensions are required on welding symbols shown at a, c and d, but not at b. Also, the note specifying STAGGERED INTERMITTENT WELD normally would appear at right of welding symbol.

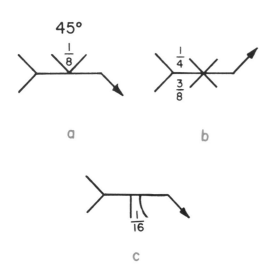

Fig. 19-7. Contour symbols.

Fig. 19-8. Groove symbols.

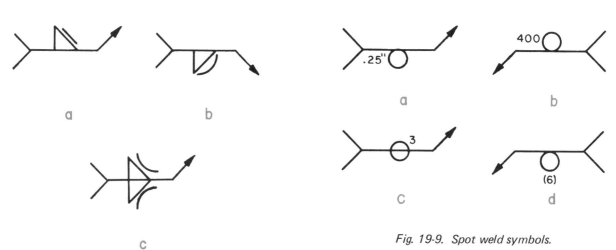

Fig. 19-9. Spot weld symbols.

Fig. 19-10. Weld-all-around symbol.

Fig. 19-11. Field weld symbol.

FIELD WELD symbol, Fig. 19-11, indicates the welds which are not made in the shop or place of initial construction.

MELT-THRU symbol indicates the welds where 100 percent joint or member penetration plus reinforcement are required in welds made from one side. See Fig. 19-12(a). When melt-thru welds are to be finished by machining or some other process, a contour symbol is added, Fig. 19-12(b).

FINISH SYMBOL indicates method of finishing (C = chipping, G = grinding, M = machining, R = rolling, H = hammering) and not the surface texture. See symbol at Fig. 19-12(b).

Fig. 19-12. Melt-thru symbol.

Blueprint Reading Activity 19−1
READING A WELDING BLUEPRINT FOR A STORAGE TANK ROOF VENT

Refer to Print 19-1 on page 184 to answer the following questions.

1. Give the name and drawing number of the assembly.

1. _____

2. What is part number 15-1, and what size is it?

2. _____
_____

3. Give the weld specification for joining parts 15-1 and 15-2.

3. _____
_____
_____

4. Give the weld specification for joining parts 15-3 and 15-4.

4. _____
_____

5. What is the size of part 15-4? Indicate the weld specification for the joint after forming.

5. _____
_____

6. How is part 15-4 to be welded to 15-5?

6. _____
_____

7. What is part 15-9? To what and how is it to be welded?

7. _____
_____
_____

8. Machine screws are used to hold the brackets supporting the head. How many brackets are there, and how are the nuts to be held in place?

8. _____
_____

9. What type weld is used to fasten part 15-8 to the brackets?

9. _____

10. Give the specification for welding the Roof Vent to the Tank.

10. _____
_____
_____

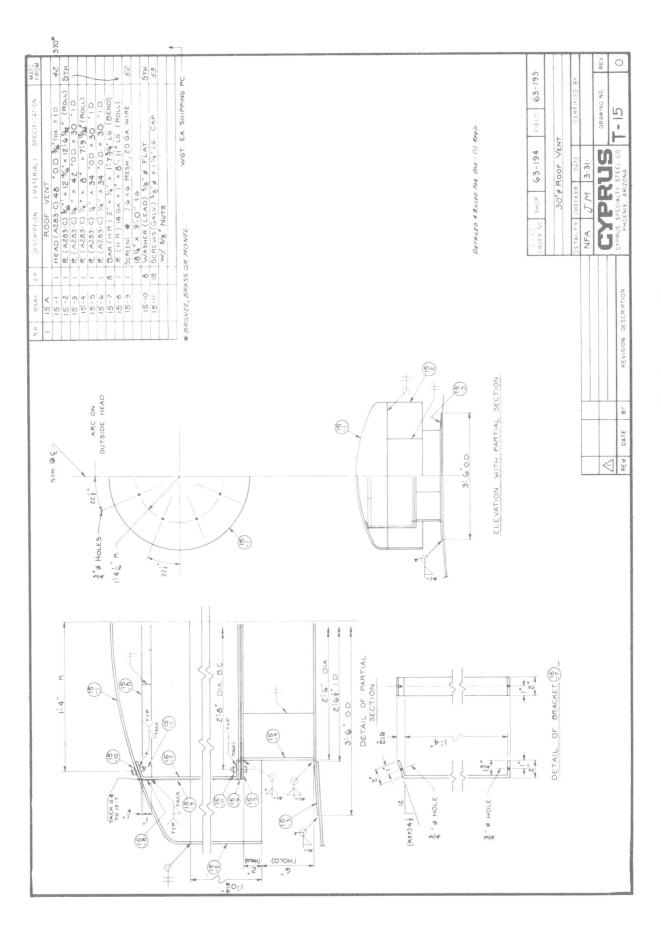

Print 19-1. Roof vent welding blueprint studied in this assignment.

Refer to Prints 19-2a through 19-2c in the Large Prints Folder to answer the following questions.

Sheet U-4

1. What is the name of the drawing?

1. _____

2. Give the weld specification for joining the lateral bracing to the rafters.

2. _____
_____
_____

3. How is the clip plate fastened to the rafter?

3. _____
_____

4. Give the weld specification for joining the rafter to the center column.

4. _____
_____

5. What is the weld specification for fastening the stiffener to the rafter?

5. _____
_____

Sheet U-5

6. Give the dimensions of the bar stock used to form part 5-6. What type weld is required?

6. _____
_____

7. What weld is required for joining parts 5-5 and 5-6?

7. _____
_____
_____

8. Give the size of the steel channels used to construct the center column.

8. _____
_____
_____
_____

9. How are parts 5-1 and 5-2 to be joined?

9. _____
_____
_____

Sheet U-6

10. What type and size beam is used for rafter 6-14?

10. _____
_____

11. What is the number of the gusset used for rafter clip 6-10? What size is this gusset?

11. _____
_____
_____

12. How are the gussets to be welded in the rafter clips?

12. _____
_____
_____

13. What is the weld specification for joining the rafter clips to the girders?

13. _____
_____

PART 4
ESTIMATING

Unit 20
Estimating Construction Costs

Estimating the costs of a construction job is one of the most important operations in the success of a contracting business. If the estimator is careless and misses items that are included in the contract or does not figure accurately, the business could suffer a severe loss. If some items are included more than once (by carelessness or by overlapping subcontractor bids), this contractor may bid too high on the job and lose it to competition.

Estimating can be a relatively simple procedure if there is a systematic approach to the task and reasonable care is given to the details of preparing the estimate of costs. In this unit, you will have an opportunity to study the general procedure used in estimating building costs and you will prepare an estimate for a typical construction job.

Methods of Estimating Costs

There are several approaches to preparing an estimate of costs for a construction project. Broadly classified, these are called the APPROXIMATE METHOD and the DETAILED METHOD. Each has a place in construction.

The APPROXIMATE METHOD of estimating costs is used by architects or contractors as follows. First, they figure the number of square or cubic feet in a construction project, then multiply this figure by the cost per unit to get an approximate cost of the job. This cost is accurate only to the following extent:

1. That the cost per unit was arrived at on a sound basis of experience.
2. That costs have not increased since the unit cost was established (or an appropriate increase is figured into the new unit cost).
3. That the construction project to be built is identical, or nearly so, to the previous one being compared.

The approximate method of estimating cost of construction is useful in helping an owner or governmental agency decide whether or not to build.

The DETAILED METHOD is the only sound basis of estimating the cost when the construction project is different in structure, size, building site, geographic area or any other aspect. The detailed method involves:

1. A careful study of the blueprints and specifications.
2. "Taking off" (listing) everything including materials, labor, allowance for materials waste, equipment costs (depreciation or rental fees), supervision, overhead and profit.

Steps in Preparing a Detailed Estimate

The following steps will assist you in developing a systematic approach to the detailed estimate of construction costs:
1. Study the construction blueprints.
2. Study the construction specifications.

3. Take off labor and materials (this usually follows the headings of the specifications).
4. Price labor and materials and calculate total costs.
5. Add other costs.
   a. Fees for permits.
   b. Utilities: electric, gas, telephone, water and sewer hookups.
   c. Insurance protection for workers and materials.
   d. Overhead for business or office operation.
   e. Profit for job.

As you enter items on the take off sheets, check these off on the blueprints and specifications.

Forms for Preparing a Detailed Estimate

Many contractors have developed a set of forms that suits their particular need in preparing an estimate. Fig. 20-1 shows a portion of a typical Estimate Take Off and Cost Sheet. For clarification, sample entries have been made on the sheet under the various headings. Extra copies of the estimate sheet appear in the Reference Section.

Checking a Detailed Estimate

Having arrived at a detailed estimate of construction costs, the estimator should carefully check through the blueprints and specifications for all items in the contract. Then, if all things have been considered, the estimate should be checked by the approximate method, preferably by another person. The cost can be roughly compared to a similar local project which has already been completed. This manner of checking will help avoid any serious loss due to omission of an item or an error in calculations.

ESTIMATE TAKE OFF AND COST SHEET

Project: _____

Item No.	Item Identification	Location on Job	Cost			Total Cost
			Labor	Material	Equipment	
02100	Site preparation	Lot No. 163	$350		$150	$500
03300	Concrete slab (Job Performed)	Driveway (Identify with Project)	$500	$325	$75	$900
	(Number assigned to an item during "take off.")					
950	Amount of bid					$185,275.00

Fig. 20-1. Form suitable for estimating the cost of construction of houses and light commercial building projects.

Blueprint Reading Activity 20−1
ESTIMATING CONSTRUCTION COSTS FOR THE
BRICK RESIDENCE

Refer to Prints 20-1a through 20-1f in the Large Prints Folder to estimate the cost of the complete residence. Use the detailed method of estimating. Include all costs except the lot for the residence, which you may assume the owner now owns.

Figure labor and materials for the entire job or figure only those jobs you will directly perform or supervise and get cost estimates on those phases you plan to subcontract. Whichever plan your instructor directs you to follow, you are responsible for the estimate of costs for the entire project.

The following outline will help you in organizing the estimate.

Outline for Preparing a Detailed Estimate

1. Prepare a cost estimate using the form "Estimate Take Off and Cost Sheet" shown in Fig. 20-2. Extra forms appear in the Reference Section.

2. Consider each of the following items, and enter your estimates on the cost estimate form.
   a. Plans and specs
   b. Building permits and fees
   c. Utilities
   d. Excavation and fill
   e. Concrete footings, foundations, floors
   f. Unit masonry
   g. Steel beams and columns
   h. Lumber, framing
   i. Lumber, finish
   j. Hardware
   k. Roofing
   l. Windows and screens
   m. Doors and screens
   n. Electric wiring
   o. Electric fixtures
   p. Sheet metal ducts and gutters
   q. Air conditioning
   r. Plumbing
   s. Insulation
   t. Gypsum board
   u. Cabinets
   v. Counter tops
   w. Tile - ceramic, vinyl
   x. Floor sanding
   y. Painting and decorating
   z. Glazing and mirrors
   aa. Floor coverings
   ab. Window cleaning
   ac. Built-ins and accessories
   ad. Range-oven
   ae. Vent hood
   af. Dishwasher
   ag. Garbage disposal
   ah. Tub-shower enclosures
   ai. Fireplace mantel
   aj. Clean up
   ak. Insurance protection

   In addition to these items, you may want to consider business overhead, finance charges and profit on the job.

3. The list of materials that follows was prepared by the Garlinghouse Plan Service. It includes all items except nails, screws, paints, calking, flashings, hardware, electrical, plumbing and air conditioning materials. Use this list as a basis of your estimate, adding the other items omitted here and items other than materials listed under step 2.

4. When you have completed your detailed estimate of costs, check your estimate by using the approximate method. Your instructor will assist you in arriving at an approximate cost per unit for your area.

ESTIMATE TAKE OFF AND COST SHEET

Project: _____

Item No.	Item Identification	Location on Job	Cost			Total Cost
			Labor	Material	Equipment	

Fig. 20-2. *Form suitable for estimating the cost of construction of the Brick Residence.*

```
M A T E R I A L S   L I S T
PLAN NO. 350
GARLINGHOUSE PLAN SERVICE
```

```
  54  4 x 8 x 16 solid concrete blocks for exterior walls
1600  Norman bricks for terrace wall
7000  Norman bricks for exterior veneer
 21½  cu. yds. concrete for basement floor slab
   7  cu. yds. concrete for garage floor slab and apron
   6  cu. yds. concrete for terrace, porch, entrance platforms and steps
  13  cu. yds. concrete for footings
  57  cu. yds. concrete for foundation walls
 200  lin. ft. 4" drain tile
```

STRUCTURAL STEEL

```
   1  8" WF 17 lb. beam 59'-1" long
   1  8" WF 17 lb. beam 26'-4" long
   5  4" steel pipe columns 7'-2" long with plates

   7  3" x 3" x 1/4" steel angle lintels 44" long over basement windows
   1  13½" x 3/8"   steel plate 20'-0 3/4" long for flitch beam
   1  Complete gas vent


3400  lin. ft. No. 4 reinforcing rods
3000  sq. ft. 6" x 6" x 8/8 gauge reinforcing mesh
   7  galvanized steel areaways 36" diameter 24" high
  70  1/2" anchor bolts 10" long
```

CARPENTER'S LUMBER

```
   5  pcs. 2 x 8 x  8 ft. long  )
   1  pc.  2 x 8 x 12 ft. long  )  First floor joists & headers
 131  pcs. 2 x 10 x 14 ft. long )

   1  pc.  2 x 6 x  8 ft. long  )
   7  pcs. 2 x 6 x 12 ft. long  )  Sub sills
  15  pcs. 2 x 6 x 14 ft. long  )

 324  lin. ft. 1" x 3"  . . . .  Cross bridging

  14  pcs. 2 x 6 x 10 ft. long  )
  75  pcs. 2 x 6 x 14 ft. long  )  House ceiling joists
  13  pcs. 2 x 6 x 16 ft. long  )

  17  pcs. 2 x 4 x 10 ft. long  )  Porch ceiling joists

   1  pc.  2 x 8 x  8 ft. long  )
  16  pcs. 2 x 8 x 22 ft. long  )  Garage ceiling joists
   8  pcs. 2 x 6 x 12 ft. long  )

  10  pcs. 1 x 6 x 8 ft. long   )  Collar beams

  14  pcs. 2 x 4 x 16 ft. long  )  Roof bracing
  24  pcs. 2 x 4 x 10 ft. long  )

   4  pcs. 2 x 12 x 14 ft. long )
   3  pcs. 2 x 12 x 12 ft. long )  Wood beams
   1  pc.  2 x 12 x 16 ft. long )

   8  pcs. 6 x 6 x 7 ft. long   )  Turned wood posts
 310  lin. ft. 2 x 4 . . . . .     Cornice nailing blocks

  66  lin. ft. 2" x 4"          )
 158  lin. ft. 2" x 12"         )  Window and door headers & sill framing
  40  lin. ft. 2" x 14"         )
```

PLAN No. 350   Page 2

```
   3   2 x 12 x 14' long ..............Carriage        )
  12   2 x 10 x 3'-6" long ...........Treads           )  Basement
   2   2" diam. x 12' long ...........Hand rail        )     stairs
   2   2 x 4 x 4' long ...............Railing posts    )

 500   pcs. 2 x 4 x 8-0 ft. long )  First floor outside wall
  11   pcs. 2 x 8 x 8-0 ft. long )  and partition studs

 125   pcs. 2 x 4 x 12 ft. long )  Head and sole plates
   3   pcs. 2 x 8 x 10 ft. long )

   8   pcs. 2 x 6 x 8 ft. long )  Rafter ties
 880   lin. ft. 2 x 4           )  Rafter ties and bearing plates

  72   pcs. 2 x 4 x 8 ft. long  )  Garage studs
  18   pcs. 2 x 4 x 12 ft.long  )  Head and sole plates

   6   pcs. 2 x 6 x 14 ft.long  )
   6   pcs. 2 x 6 x 16 ft.long  )  Common and jack rafters
  81   pcs. 2 x 6 x 20 ft.long  )
  56   pcs. 2 x 6 x 22 ft.long  )

   4   pcs. 2 x 8 x 26 ft. long )  Hip rafters
   2   pcs. 2 x 8 x 26 ft. long )

   2   pcs. 2 x 8 x 26 ft. long )  Valley rafters

   1   pc.  2 x 8 x 12 ft. long )
   2   pcs. 2 x 8 x 14 ft. long )  Ridge boards
   1   pc.  2 x 8 x 16 ft. long )
```

```
1920   sq. ft. 5/8" plyscord plywood - first floor sub flooring
1760   sq. ft. 3/8" plyscord plywood - wall sheathing
4160   sq. ft. 3/8" plyscord plywood - roof sheathing
3680   sq. ft. 15 lb. asphalt felt, flooring paper & sheathing paper
  40   squares 18" x 3/8" to 3/4" handsplit-resawn Red Cedar shakes-roofing
                                                                 material
1900   sq. ft. Polyethelene- vapor barrier
8000   sq. ft. 30 lb. roofing felt
1216   sq. ft. 1/2" gypsum wallboard - garage walls and ceiling finish
 192   sq. ft. 3/8" plywood - porch ceiling
   8   1'-6" x 3'-10" - shutters
 250   lin. ft. frieze molding
 512   sq. ft. 3/8" plywood - soffit
 115   lin. ft. 1 x 6 beam casing
 278   lin. ft. 1 x 2 fascia trim
 278   lin. ft. 1 x 8 fascia
```

INTERIOR WALL & CEILING FINISH
```
6784   sq. ft. 1/2" gypsum wallboard
```

MILLWORK
DOORS:
```
   2   outside door frames, openings, 5-0 x 6-8, jambs 1-5/16" x 4-3/8"
   1   outside door frame,  opening,  2-8 x 6-8, jambs 1-5/16" x 4-3/8"
   1   outside door frame,  opening,  2-8 x 6-8, jambs 1-5/16" x 4-1/2"
   1   inside  door frame,  opening,  6-0 x 6-8, jambs 3/4" x 4-1/2"
   1   inside  door frame,  opening,  5-0 x 6-8, jambs 3/4" x 4-1/2"
   5   inside  door frames, openings, 4-0 x 6-8, jambs 3/4" x 4-1/2"
   2   inside  door frames, openings, 3-0 x 6-8, jambs 3/4" x 4-1/2"
   2   inside  door frames, openings, 2-8 x 6-8, jambs 3/4" x 4-1/2" with stops
   3   inside  door frames, openings, 2-6 x 6-8, jambs 3/4" x 4-1/2" with stops
   5   inside  door frames, openings, 2-0 x 6-8, jambs 3/4" x 4-1/2" with stops
   1   recessed door frame, opening,  2-6 x 6-8
   1   recessed door frame, opening,  2-0 x 6-8
   1   garage   door frame, opening, 18-0 x 7-0, jambs 3/4" x 5-3/8" with stops

   4   outside entrance doors      2-6 x 6-8 x 1-3/4
   1   outside service door        2-8 x 6-8 x 1-3/4
```

```
  2  inside doors          3-0 x 6-8 x 1-3/8
  3  inside doors          2-8 x 6-8 x 1-3/8
  6  inside doors          2-6 x 6-8 x 1-3/8
 16  inside doors          2-0 x 6-8 x 1-3/8
  1  garage door          18-0 x 7-0 x 1-3/8
  1  combination door      2-8 x 6-8 x 1-1/8
  4  combination doors     2-6 x 6-8 x 1-1/8

  2  sides of door trim    6-0 x 6-8, 3/4 x 2½
  4  sides of door trim    5-0 x 6-8, 3/4 x 2½
 10  sides of door trim    4-0 x 6-8, 3/4 x 2½
  4  sides of door trim    3-0 x 6-8, 3/4 x 2½
  7  sides of door trim    2-8 x 6-8, 3/4 x 2½
  8  sides of door trim    2-6 x 6-8, 3/4 x 2½
 12  sides of door trim    2-0 x 6-8, 3/4 x 2½
```

WINDOWS:
All*Andersen windows are to be complete with frames, sash, interior
trim, exterior trim, screens, storm sash and hardware.

```
  3  No. 24-2432-24    Andersen Narroline D.H. window units
  3  No. 2832-28       Andersen Narroline D.H. window units
  1  No. 2432-24       Andersen Narroline D.H. window unit
  2  No. 2832          Andersen Narroline D.H. windows
  3  No. 2432          Andersen Narroline D.H. windows
     *Andersen Corporation, Bayport, Minnesota

  7  alum. basement units, 2 lites 15" x 20" with screens
```

CABINETS AND MISCELLANEOUS MILLWORK

```
14'-3" kitchen base unit, 36" high 24" deep, complete with hardware
 6'-9" kitchen wall unit, 30" high 12" deep, complete with hardware
 5'-9" kitchen wall unit, 18" high 12" deep, complete with hardware
 2'-3" kitchen oven unit, 84  high 24" deep, complete with hardware

  1  80" wide 22" deep 31" high, lavatory cabinet complete
  1  75½" & 62" wide 22" deep 34" high, lavatory cabinet complete
  1  30" wide 16" deep 84" high, towel cabinet complete
  1  36" wide 24" deep 29" high, desk complete

 352  sq. ft. 5/8" (C-REP.D) plywood - flooring under vinyl
1750  BM 25"/32 x 2½" clear oak T. & G. end matched - finished flooring

  70  lin. ft. 1" x 12" closet shelving
 320  sq. ft. vinyl flooring
  50  sq. ft. slate tile
 540  lin. ft. 9/16" x 3" base mold
 540  lin. ft. 1/2 x 3/4" base shoe

1820  sq. ft. 6"    thick ceiling insulation
1300  sq. ft. 3-1/2" thick wall insulation
```

MISC. ITEMS

```
  44  lin. ft. 1-3/8" closet poles
```

While every attempt has been made in the preparation of this material
list to avoid mistakes, the maker cannot guarantee against human errors.
The contractor must check all quantities, etc. and be responsible for same.
Due to the different practices of contractors, variation in climatic
conditions, different covering capacities of paints, wide range of hardware
requirements and various local building codes and practices, no attempt
has been made to figure items such as nails, screws, paint, calking,
flashing, hardware or electrical, plumbing and heating materials.

Blueprint Reading Activity 20−2
ESTIMATING CONSTRUCTION COSTS FOR THE TUDOR DUPLEX

Refer to Prints 20-2a through 20-2e in the Large Prints Folder to estimate the cost of the complete residence. Use the detailed method of estimating. Record all costs, including a suitable lot for the duplex which you may select and obtain the price estimate.

Figure labor and materials for the entire job or figure only those jobs you will directly perform or supervise and get cost estimates on those phases you plan to subcontract. Whichever plan your instructor directs you to follow, you are responsible for the estimate of costs for the entire project.

The following outline will help you in organizing the estimate.

Outline for Preparing a Detailed Estimate

1. Prepare a cost estimate sheet using the form "Estimate Take Off and Cost Sheet" shown in Fig. 20-3. Extra forms appear in the Reference Section.

2. Consider each of the following items, and enter your estimates on the cost estimate form.
   a. Plans and specs
   b. Building permits and fees
   c. Utilities
   d. Excavation and fill
   e. Conc footings, foundations, floors
   f. Unit masonry
   g. Steel beams and columns
   h. Lumber, framing
   i. Lumber, finish
   j. Hardware
   k. Roofing
   l. Windows and screens
   m. Doors and screens
   n. Electric wiring
   o. Electric fixtures
   p. Sheet metal ducts and gutters
   q. Air conditioning
   r. Plumbing
   s. Insulation
   t. Gypsum board
   u. Cabinets
   v. Counter tops
   w. Tile - ceramic, vinyl
   x. Floor sanding
   y. Painting and decorating
   z. Glazing and mirrors
   aa. Floor coverings
   ab. Window cleaning
   ac. Built-ins and accessories
   ad. Range-oven
   ae. Vent hood
   af. Dishwasher
   ag. Garbage disposal
   ah. Tub-shower enclosures
   ai. Fireplace mantel
   aj. Clean up
   ak. Insurance protection

   In addition to these items, you may want to consider business overhead, finance charges and profit on the job.

3. When you have completed your detailed estimate of costs, check your estimate by using the approximate method. Your instructor will assist you in arriving at an approximate cost per unit for your area.

ESTIMATE TAKE OFF AND COST SHEET

Project:_____

Item No.	Item Identification	Location on Job	Cost			Total Cost
			Labor	Material	Equipment	

Fig. 20-3. Form suitable for estimating the cost of construction of the Tudor Duplex.

PART 5
ADVANCED BLUEPRINT READING
SETS OF BLUEPRINTS

Unit 21
Reading a Set of Blueprints
for a Residence

In the preceding units, you have been reading blueprints for a single phase of construction, such as footings and foundations or framing. In this unit, the architectural rendering, Fig. 21-1, cost estimate form, Fig. 21-2, and the set of seven blueprints, Prints 21-1a through g of the residence are typical of those you would use in constructing most houses.

There are four assignments over the full set of prints: three are on reading the construction drawings; one is on estimating the costs. In looking for answers to the questions in this unit, you will be working back and forth from one print to another, just as you would on a construction job. Good luck in handling these assignments.

Fig. 21-1. A sketch of the contemporary residence studied in this unit. (Garlinghouse Plan Service)

Refer to Prints 21-1a through 21-1g in the Large Prints Folder to answer the following questions.

1. What is the scale of the foundation plan?

1. _____

2. Give overall length and width dimensions of foundation through the house in line with:
   a. The family room - bedroom
   b. The family room - garage
   c. The master bedroom - front bedroom

2. a. _____
   b. _____
   c. _____

3. What size footing and what reinforcing steel are required at A?

3. _____
   _____

4. Give the size of the footings at B, C, D.

4. B _____
   C _____
   D _____

5. What are the dimensions of the fireplace footings, and what reinforcing steel is required?

5. _____
   _____
   _____

6. How are the footings and piers that support the beam below the bedroom to be constructed?

6. _____
   _____
   _____
   _____

7. To what depth are the footings to be placed at D, E, F?

7. D _____
   E _____
   F _____

8. Explain how the foundation walls at A and in front of the bedrooms differ from the other foundation walls. Why do they differ?

8. _____
   _____
   _____
   _____

9. How does the height of the foundation wall at the garage door opening differ from the adjacent wall?

9. _____

10. What special forming is necessary in the foundation walls for the beams below:
    a. The kitchen - living room?
    b. The bedrooms?

10. a. _____
    b. _____

11. What concrete slabs are to be poured? How are they reinforced?

11. _____
    _____

Advanced Blueprint Reading Activity 21—2
READING THE FRAMING AND UNIT MASONRY BLUEPRINTS
FOR THE CONTEMPORARY RESIDENCE

Refer to Prints 21-1a through 21-1g in the Large Prints Folder to answer the following questions.

1. Give the scale of the drawings for:
   a. The floor plan
   b. The elevations
   c. The clerestory window detail

1. a. _____
   b. _____
   c. _____

2. Give overall length and width dimensions of frame walls through the house in line with:
   a. The family room - bedroom
   b. The family room - garage
   c. The master bedroom - front bedroom

2. a. _____
   b. _____
   c. _____

3. What size beam is called for in the bedroom area?

3. _____
   _____

4. What size beams are specified in the living room - kitchen area? How are they supported?

4. _____
   _____
   _____
   _____
   _____

5. Floor joists of 2x10s are specified generally over the floor area and beams. How does the print show that the same size is required for the shorter joists in front of the fireplace opening?

5. _____
   _____

6. How many rows of bridging are required?

6. _____

7. How are the double joists below the partition walls in the bedroom area to be constructed?

7. _____
   _____

8. What is the size of the rough opening for the fireplace and hearth?

8. _____

9. Give the wall stud size and spacing in the exterior wall.

9. _____

10. Describe the exterior wall construction at H.

10. _____
    _____
    _____
    _____
    _____
    _____
    _____
    _____

11. How is the exterior wall at J to be constructed?

11. _____

12. How is the common wall between the two baths to be constructed?

12. _____

13. On what sheet is the detail of the construction of the entry door wall?

13. _____

14. Give the rough opening size for doors.

14. A _____
    B _____
    E _____
    F _____

15. What is the rough opening size for the garage door? What size header is required?

15. _____

16. Give the rough opening size for the windows at K, L.

16. K _____
    L _____

17. How far off the corner stud face does window V center? What kind of a window is it?

17. _____

18. Indicate the number and size of the following for the stairway:
    a. Carriages
    b. Risers
    c. Treads

18. a. _____
    b. _____
    c. _____

19. What material is used for the fireplace:
    a. Foundation wall
    b. Finish siding
    c. Damper

19. a. _____
    b. _____
    c. _____

20. How is the fireplace hearth to be constructed?

20. _____

21. What size are the rafter joists, and how are they to be spaced?

21. _____

22. What size and how long are the beams over:
    a. The foyer
    b. The kitchen
    c. The bedrooms (3 beams)
    d. The portico (2 beams)
    e. The garage (3 beams)

22. a. _____
    b. _____
    c. _____
    d. _____
    e. _____

23. Give the size of the roof opening for the chimney.

23. _____

24. What material is used to form the roof?

24. _____
_____

25. What insulation is specified for the walls and roof-ceiling?

25. _____

26. What material is used for the soffit, fascia and trim of the eaves?

26. _____

27. The material used for the interior wall and ceiling finish is:

27. _____

28. What is the kitchen counter height?

28. _____

29. What are the dimensions of the kitchen sink cabinet on each side at the back and on front?

29. _____

30. How are the backs of the counter top stove cabinets to be finished?

30. _____

Refer to Prints 21-1a through 21-1g in the Large Prints Folder to answer the following questions.

1. How many lavatories are shown on the plans?

1. _____

2. Give the number of the following fixtures shown on the plan:
   a. Bath tub
   b. Shower stall
   c. Water closet
   d. Hose bibb
   e. Hot water heater

2. a. _____
   b. _____
   c. _____
   d. _____
   e. _____

3. What specifications are given for the kitchen sink?

3. _____
   _____

4. Where is gas to be plumbed?

4. _____
   _____

5. How many of the following electrical fixtures are shown on the plans?
   a. Ceiling light
   b. Pull chain light
   c. Exterior light
   d. Fluorescent ceiling light
   e. Fluorescent wall bracket
   f. Concealed soffit light

5. a. _____
   b. _____
   c. _____
   d. _____
   e. _____
   f. _____

6. List the number of the following devices shown on the plans:
   a. Single pole switch
   b. 3-way switch
   c. 4-way switch
   d. Duplex conv. outlet
   e. Range, oven dryer outlet
   f. Telephone outlet

6. a. _____
   b. _____
   c. _____
   d. _____
   e. _____
   f. _____

Reading a Set of Blueprints for a Residence

ESTIMATING THE CONSTRUCTION COSTS
FOR THE CONTEMPOERARY RESIDENCE

Refer to Prints 21-1a through 21-g in the Large Prints Folder to estimate the costs of the complete construction project using the detailed method of estimating. You should include all costs, except the lot for the residence which you may assume the owner now owns.

Figure labor and materials for the entire job or figure only those jobs you will directly perform or supervise and get cost estimates on those phases you plan to subcontract. Whichever plan your instructor directs you to follow, you are responsible for the estimate of costs for the entire program.

The following outline will help you in organizing the estimate.

Outline for Preparing a Detailed Estimate

1. Prepare a cost estimate using the form "Estimate Take Off and Cost Sheet" shown in Fig. 21-2. Extra copies of the form appear in the Reference Section. This form was discussed in Unit 20.

2. Consider each of the following items and enter your estimates on the cost estimate form.
   a. Plans and specs
   b. Building permits and fees
   c. Utilities
   d. Excavation and fill
   e. Conc footings, foundations, floors
   f. Unit masonry
   g. Steel beams and columns
   h. Lumber, framing
   i. Lumber, finish
   j. Hardware
   k. Roofing
   l. Windows and screens
   m. Doors and screens
   n. Electric wiring
   o. Electric fixtures
   p. Sheet metal ducts and gutters
   q. Air conditioning
   r. Plumbing
   s. Insulation
   t. Gypsum board
   u. Cabinets
   v. Counter tops
   w. Tile - ceramic, vinyl
   x. Floor sanding
   y. Painting and decorating
   z. Glazing and mirrors
   aa. Floor coverings
   ab. Window cleaning
   ac. Built-ins and accessories
   ad. Range-oven
   ae. Vent hood
   af. Dishwasher
   ag. Garbage disposal
   ah. Tub-shower enclosures
   ai. Fireplace mantel
   aj. Clean up
   ak. Insurance protection

   In addition to these items, you may want to consider business overhead, finance charges and profit on the job.

3. When you have completed your detailed estimate of costs, check your estimate by using the approximate method. Your instructor will assist you in arriving at an approximate cost per unit for your area.

ESTIMATE TAKE OFF AND COST SHEET

Project: _____

Item No.	Item Identification	Location on Job	Cost			Total Cost
			Labor	Material	Equipment	

Fig. 21-2. Form suitable for estimating the cost of construction of the Contemporary Residence.

Unit 22
Reading a Set of Blueprints for a Fourplex

In this unit, you will have the opportunity to study the architectural rendering, Fig. 22-1, and four blueprints for a multiple unit dwelling. There are four assignments over the set of prints: three are on reading the Fourplex construction drawings; one is on estimating the costs. You will be referring to the full set of prints, just as you would on the construction job.

Fig. 22-1. A sketch of the Fourplex studied in this unit.

Advanced Blueprint Reading Activity 22—1
READING THE FOOTING, FOUNDATION AND CONCRETE
FLOOR BLUEPRINTS FOR THE FOURPLEX

Refer to Prints 22-1a through 22-1d in the Large Prints Folder to answer the following questions.

1. What is the scale of the foundation plan?

1. _____

2. Give the overall length and width dimensions of the building.

2. _____
   _____

3. What size footing and what reinforcing steel are required in the wall footing at A, B, C?

3. A _____
   _____
   _____
   B _____
   C _____

4. How deep is the wall footing to be placed?

4. _____

5. Describe the foundation wall at A, B, C.

5. A _____
   _____
   _____
   _____
   B _____
   C _____
   _____

6. How are the corners of the foundation wall to be reinforced?

6. _____
   _____
   _____
   _____

7. What size opening for the beam is to be provided in the exterior foundation wall? Interior wall?

7. Exterior: _____
   Interior: _____

8. Give the size of the door to the crawl space.

8. _____
   _____

9. How many piers are there? What are their requirements?

9. _____
   _____
   _____

10. What clearance is to be provided under the floor of the building?

10. _____

11. How is the crawl space vented and treated?

11. _____
_____
_____
_____
_____
_____

12. What is to be done with the ground beneath the porches?

12. _____
_____
_____
_____

Advanced Blueprint Reading Activity 22—2
READING FRAMING AND UNIT MASONRY BLUEPRINTS
FOR THE FOURPLEX

Refer to Prints 22-1a through 22-1d in the Large Prints Folder to answer the following questions.

1. What is the scale of the floor plan?

1. _____

2. How are the two exterior side walls to be constructed?

2. _____
_____
_____
_____
_____

3. What type and size beam is specified for support beneath the first floor joists?

3. _____
_____

4. What size sill is required above the foundation wall and beam?

4. _____

5. Give the size of the floor joists and indicate their spacing?

5. _____

6. How are the joists to be framed under D?

6. _____
_____

7. What is the subfloor material?

7. _____

8. What type of framing — balloon or platform — is called for on the blueprints?

8. _____

9. What size studs are called for at E? How are they to be spaced?

9. _____

10. What type of a wall is indicated at F?

10. _____

11. How is the wall to be framed at G?

11. _____

12. What is the distance from the inside of the exterior wall on the right to the center of wall at D, E, H?

12. D _____
E _____
H _____

13. Give the dimensions of the utility rooms, wall center to wall center for the first floor? Second floor?

13. First: _____
Second: _____

14. How do the second floor beams support the balcony floor joists?

14. _____
_____

15. What is the size of the beam in question 14?

15. _____

16. How are the balcony floor joists cut and framed?

16. _____

17. What bridging is required in the floor joists?

17. _____

18. Give the number and size of the following for the stairway framing:
    a. Carriages
    b. Risers
    c. Treads
    d. Stringer

18. a. _____
    b. _____
    c. _____
    d. _____

19. What material is to be used in the exterior wall above the stairway?

19. _____

20. How many roof trusses are to be used? What size are they?

20. _____

21. What sheathing is indicated for:
    a. The exterior wall, front and rear
    b. The roof

21. a. _____
    b. _____

22. Describe the roof material.

22. _____

23. What is to be used for the interior wall covering?

23. _____

24. What flooring is to be used in:
    a. Living room
    b. Kitchen
    c. Bath
    d. Bedroom

24. a. _____
    b. _____
    c. _____
    d. _____

25. Furring down for the kitchen cabinets is at what height?

25. _____

Advanced Blueprint Reading Activity 22—3
READING THE PLUMBING AND ELECTRICAL BLUEPRINTS
FOR THE FOURPLEX

Refer to Prints 22-1a to 22-1d in the Large Prints Folder to answer the following questions.

1. How many lavatories are shown on the plans for the four apartments?

1. _____

2. What type of kitchen sink is shown?

2. _____

3. Where is the water heater located?

3. _____

4. Assume the water main is available in the street in front of the building. Sketch a water supply distribution diagram for apartment A with a separate meter.

4.

5. Sketch an isometric diagram of a sewage disposal system for the four apartments. Assume the public sewer is to the rear of the building.

5.

6. How many of these electrical fixtures are shown on the plans for the four apartments?
   a. Ceiling light
   b. Pull chain light
   c. Ceiling lighter-timer sw.
   d. Exterior light fixture
   e. Fluorescent ceiling light
   f. Fluorescent wall light

6. a. _____
   b. _____
   c. _____
   d. _____
   e. _____
   f. _____

7. List the number of these devices shown on the plans for the four apartments.
   a. Single pole switch
   b. 3-way switch
   c. Duplex conv. outlet
   d. Weatherproof outlet
   e. Range, oven, dryer outlet
   f. Telephone outlet

7. a. _____
   b. _____
   c. _____
   d. _____
   e. _____
   f. _____

Advanced Blueprint Reading Activity 22−4
ESTIMATING THE CONSTRUCTION COSTS
FOR THE FOURPLEX

Refer to Prints 22-1a through 22-1d in the Large Prints Folder to estimate the costs of the complete construction project using the detailed method of estimating. You should include all costs, except the lot for the Fourplex which you may assume the owner now owns.

Figure labor and materials for the entire job or figure only those jobs you will directly perform or supervise and get cost estimates on those phases you plan to subcontract. Whichever plan your instructor directs you to follow, you are responsible for the estimate of costs for the entire project.

The following outline will help you in organizing the estimate.

Outline for Preparing a Detailed Estimate

1. Prepare a cost estimate using the form "Estimate Take Off and Cost Sheet" shown in Fig. 22-2. Extra copies appear in the Reference Section. This form was discussed in Unit 20.

2. Consider each of the following items and enter your estimates on the cost estimate form.
   a. Plans and specs
   b. Building permits and fees
   c. Utilities
   d. Excavation and fill
   e. Conc footings, foundations, floors
   f. Unit masonry
   g. Steel beams and columns
   h. Lumber, framing
   i. Lumber, finish
   j. Hardware
   k. Roofing
   l. Windows and screens
   m. Doors and screens
   n. Electric wiring
   o. Electric fixtures
   p. Sheet metal ducts and gutters
   q. Air conditioning
   r. Plumbing
   s. Insulation
   t. Gypsum board
   u. Cabinets
   v. Counter tops
   w. Tile - ceramic, vinyl
   x. Floor sanding
   y. Painting and decorating
   z. Glazing and mirrors
   aa. Floor coverings
   ab. Window cleaning
   ac. Built-ins and accessories
   ad. Range-oven
   ae. Vent hood
   af. Dishwasher
   ag. Garbage disposal
   ah. Tub-shower enclosure
   ai. Clean up
   aj. Insurance protection.
   In addition to these items, you may want to consider business overhead, finance charges and profit on the job.

3. When you have completed your detailed estimate of costs, check your estimate by using the approximate method. Your instructor will assist you in arriving at an approximate cost per unit for your area.

ESTIMATE TAKE OFF AND COST SHEET

Project:_____

Item No.	Item Identification	Location on Job	Cost			Total Cost
			Labor	Material	Equipment	

Fig. 22-2. Form suitable for estimating the cost of construction of Multiple Unit Dwellings.

Unit 23
Reading a Set of Blueprints
for a Fire Station

A set of 14 blueprints for the Fire Station are used with the assignments. This set of blueprints is typical of those for light commercial construction. By completing the assignments in this unit, you will gain experience in reading a full set of blueprints for a commercial structure.

Fig. 23-1. The blueprints for the Fire Station shown here are studied in this unit. (Charles O. Biggs, AIA)

Advanced Blueprint Reading Activity 23−1
READING THE SITE PLAN, FOUNDATION AND FLOOR PLAN BLUEPRINTS
FOR THE FIRE STATION

Refer to Prints 23-1a through 23-1n in the Large Prints Folder to answer the following questions.

1. What is the scale of the Roof and Site Plan?

1. _____

2. What restrictions are placed along the northern border of the property?

2. _____
   _____

3. Give the bearing and length of the property lines along boundaries:

3. North _____
   East _____
   South at drive _____
   West along drive _____
   South along building _____
   _____
   West along 99th Ave. _____
   _____

4. List the elevation at:
   a. SE corner of property
   b. NE corner of property
   c. NE parking lot floor sink (drain)
   d. NW parking lot floor sink (drain)
   e. Center of south drive at asphalt
   f. Center of west drive at asphalt

4. a. Existing _____ Finish _____
   b. Existing _____ Finish _____
   c. _____
   d. _____
   e. _____
   f. _____

5. What is the length and width of the building?

5. _____

6. How far is the building to be located from the property on the south? How far east of the corner on the southwest?

6. _____
   _____

7. What type of paving and underlayment are specified for the drives and parking lot?

7. _____
   _____

8. What is the scale of:
   a. The foundation plan?
   b. The details on Sheet 2?

8. a. _____
   b. _____

9. Describe the footing at: A, B.

9. A _____
   _____
   _____
   B _____
   _____
   _____

10. How much is the foundation wall to be depressed at the doorways?

10. _____

11. Describe the concrete floor in the Training Room.

11. _____
_____

12. What is to be done to this floor at the entry in lavatory-toilet room?

12. _____

13. Describe the concrete floor in the Apparatus Room.

13. _____
_____
_____
_____

14. How deep is the mechanical pit? At what elevation is the grate in the bottom?

14. _____
_____

15. What reinforcement is called for in the mechanical pit?

15. _____
_____

16. Describe the footing at C.

16. _____
_____

17. Give the dimensions of the edge of the concrete at the overhead doors.

17. _____

18. What is the overall size of the foundation plan for the Hose Drying Rack?

18. _____

19. How thick is the concrete slab surrounding the Hose Drying Rack, and at what elevation is the slab edge?

19. _____

20. What size are the weep holes in the wall at the southwest corner of the building?

20. _____

Refer to Prints 23-1a through 23-1n in the Large Prints Folder to answer the following questions.

1. What is the scale of the longitudinal and transverse sections?

   1. _____

2. What type of a section is section: AA, BB, CC?

   2. AA _____
   BB _____
   CC _____

3. Locate the following walls from the nearest exterior wall: D, E, F, G.

   3. D _____
   E _____
   F _____
   G _____

4. The width dimension for Hall #2 is _____ and this is from _____ to _____ .

   4. _____
   _____

5. The Alarm Room length dimension is _____ and this is from _____ to _____ .

   5. _____
   _____

6. The Alarm Room width dimension is _____ and this is from _____ to _____ .

   6. _____
   _____

7. The Kitchen size layout dimensions are:

   7. _____

8. The Apparatus Room length dimension is _____. The width dimension is _____, which is from _____ to _____ .

   8. _____
   _____

9. What is the masonry opening specified for the overhead doors?

   9. _____

10. Describe the wall construction at: A, B.

   10. A _____
   _____
   _____
   _____
   _____
   _____
   B _____
   _____
   _____
   _____

11. What type and size lintels are required over the doorways into the Alarm Room?

11. _____

12. Give the size and construction details of the header above the overhead doors in the Apparatus Room.

12. _____
_____
_____
_____
_____

13. What type of roof framing is called for over the Dining, Kitchen and Hall area?

13. _____
_____
_____

14. What is the slope of the above roof section?

14. _____

15. Describe the roof framing above the Captain's and Alarm Rooms.

15. _____

16. What roof joists are used above the Apparatus Room?

16. _____
_____

17. Describe the roof framing on the north and south ends of the Apparatus Room.

17. _____
_____
_____
_____
_____
_____
_____
_____

18. How is the roof structure secured to the walls at the Apparatus Room?

18. _____
_____
_____
_____

19. What type roofing material is required for the Apparatus Room roof?

19. _____

20. What type roofing material is called for above the Kitchen-Dining area?

20. _____
_____

21. What is the exterior finish material?

21. _____

22. Describe the ceiling materials and give the ceiling height in:
  a. The Training Room

22. a. _____
_____

  b. The Apparatus Room

b. _____
_____

23. How are the following areas of the Dining Room to be finished?

a. Walls

23. a. _____
_____
_____
_____

b. Floors

b. _____
_____
_____
_____

24. Describe door No. 2.

24. _____
_____

25. Describe the window in the Dining Room.

25. _____
_____
_____
_____
_____
_____

Advanced Blueprint Reading Activity 23—3
READING PLUMBING, ELECTRICAL AND AIR CONDITIONING
BLUEPRINTS FOR THE FIRE STATION

Refer to Prints 23-1a through 23-1n in the Large Prints Folder to answer the following questions.

Plumbing

1. What is the size of the building sewer and what material is to be used?

1. _____

2. Give the size of the building drain from:
   a. Kitchen
   b. Shower

2. a. _____
   b. _____

3. How does the roof drain and the two Apparatus Room drains (AD) drain from the building? See Sheets 1, 9 and 10.

3. _____
   _____
   _____

4. What other drains are channeled into this pipe?

4. _____
   _____
   _____

5. Give the size of the cleanout in:
   a. Toilet area
   b. Shower

5. a. _____
   b. _____

6. What size vent pipe is to be included for:
   a. Kitchen
   b. Lavatory-Urinal
   c. Toilet Room
   d. Rest Room

6. a. _____
   b. _____
   c. _____
   d. _____

7. Where do the vents in question 6 exit the building?

7. _____
   _____
   _____
   _____

8. What size water supply main is to come into the building?

8. _____

9. Give the size of the distribution pipe supplying cold water to:
   a. Evaporative coolers
   b. 120 gallon water heater
   c. 30 gallon water heater
   d. Showers
   e. Kitchen sink
   f. Hose bibb in hose drying area
   g. Hose bibb at Drafting Pit
   h. Hose bibb at north end of building

9. a. _____
   b. _____
   c. _____
   d. _____
   e. _____
   f. _____
   g. _____
   h. _____

10. Indicate what size hot water line serves:
    a. The kitchen sink
    b. The showers
    c. The lavatories

10. a. _____
     b. _____
     c. _____

11. Give the specifications for the water closets.

11. _____
_____
_____
_____
_____
_____

12. What sink is specified for the Kitchen?

12. _____
_____
_____

13. List the specifications for the 120 gallon water heater.

13. _____
_____
_____
_____

14. Give the specifications for the shower fixtures.

14. _____
_____

Electrical

15. What size conduit is to be stubbed out for the electric power company? How far?

15. _____

16. How is the electric power system to be grounded?

16. _____
_____
_____
_____

17. Interpret the specifications for the Service Entrance Section.

17. _____
_____
_____
_____
_____
_____
_____
_____

18. Which water heater is supplied current directly from the Service Entrance Section and what size conductor and conduit is to be used?

18. _____

19. How many and what size conductors and conduit are to be used for the feeder circuit to Panel A? Panel EM?

19. A _____
     EM _____

20. Interpret information at H on Sheet 12.

20. _____

21. Interpret information at J on Sheet 12. What size conduit is required for the circuit?

21. _____

22. Explain information given at K on Sheet 12.

22. _____

23. Where is the telephone mounting board to be located?

23. _____

24. Interpret information given at L on Sheet 12.

24. _____

25. Interpret information given at M on Sheet 13.

25. _____

26. Where are Panels A and EM located?

26. A _____
    EM _____

27. What type of lighting fixture is specified for the ceiling lights in the Dormitory?

27. _____

28. Give the specifications for the lighting fixtures in the Service Pit.

28. _____

29. At what height are the receptacles in the Apparatus Room to be mounted?

29. _____

30. Which lights in the Apparatus Room are operable under emergency power.

30. _____

Air Conditioning

31. Give the specifications for the air conditioning unit.

31. _____

32. The A/C units are to be mounted on what and at what height?

32. _____

33. What size supply and return air ducts lead to and from the A/C units?

33. _____

34. Give the size of the ducts in the Training Room.

34. _____

35. What size ceiling diffusers are to be installed in the Training Room?

35. _____

36. Give the duct and diffuser size in the Dining Room.

36. _____

37. How is the Apparatus Room cooled? Where are these located?

37. _____

38. Give the specifications for the evaporative coolers.

38. _____

39. What piece of equipment is at N on Sheet 14 and how does it work?

39. _____

40. Where is exhaust fan No. 2 located and where does it discharge? Give its specifications.

40. _____

Advanced Blueprint Reading Activity 23—4
ESTIMATING THE CONSTRUCTION COSTS FOR THE FIRE STATION

Refer to Prints 23-1a through 23-1n in the Large Prints Folder to estimate the costs of the complete construction project using the detailed method of estimating. You should include all costs, except for the lot which you may assume the city owns.

Figure labor and materials for the entire job or figure only those jobs you will directly perform or supervise and get cost estimates on those phases you plan to subcontract. Whichever plan your instructor directs you to follow, you are responsible for the estimates of costs for the entire project.

The following outline will help you in organizing the estimate.

Outline for Preparing a Detailed Estimate

1. Prepare a cost estimate using the form "Estimate Take Off and Cost Sheet" shown in Fig. 23-2. Extra copies of the form appear in the Reference Section. This form was discussed in Unit 20.

2. Consider each of the following items, and enter your estimate on the cost estimate form.
   a. Plans and specs
   b. Building permits and fees
   c. Utilities
   d. Excavation and fill
   e. Conc footings, foundations, floors
   f. Unit masonry
   g. Steel beams and columns
   h. Lumber, framing
   i. Lumber, finish
   j. Hardware
   k. Roofing
   l. Windows and screens
   m. Doors and screens
   n. Electric wiring
   o. Electric fixtures
   p. Sheet metal ducts and gutters
   q. Air conditioning
   r. Plumbing
   s. Insulation
   t. Gypsum board
   u. Cabinets
   v. Counter tops
   w. Tile - ceramic, vinyl
   x. Floor sanding
   y. Painting and decorating
   z. Glazing and mirrors
   aa. Floor coverings
   ab. Window cleaning
   ac. Built-ins and accessories
   ad. Range-oven
   ae. Vent hood
   af. Dishwasher
   ag. Garbage disposal
   ah. Tub-shower enclosure
   ai. Clean up
   aj. Insurance protection
   In addition to these items, consider business overhead, finance charges and profit on the job.

3. When you have completed your detailed estimate of costs, check your estimate by using the approximate method. Your instructor will assist you in arriving at an approximate cost per unit for your area.

ESTIMATE TAKE OFF AND COST SHEET

Project:_____

Item No.	Item Identification	Location on Job	Cost			Total Cost
			Labor	Material	Equipment	

Fig. 23-2. Form suitable for estimating the cost of construction of the Fire Station.

Unit 24
Reading a Set of Blueprints
for an Office Building

A set of 17 blueprints for the Office Building are used with the three blueprint reading assignments. In Assignments 24-1 and 24-2, the series of questions can be answered only through careful examination of the prints that relate. In some cases, more than one print is involved in finding the answers.

In Assignment 24-3, you will be asked to make an in-depth study of the blueprints and do a detailed estimate of construction costs.

These assignments will provide you with additional experience in reading a full set of prints for a commercial construction job.

Fig. 24-1. A sketch of the Office Building studied in this unit. (Smith & Neubek, Architects)

Advanced Blueprint Reading Activity 24−1
READING THE FOUNDATION, FRAMING AND INTERIOR FINISH BLUEPRINTS
FOR THE OFFICE BUILDING

Refer to Prints 24-1a through 24-1q in the Large Prints Folder to answer the following questions.

1. What is the scale of the plot plan?

    1. _____

2. Give the overall size of the plot, including the part for the future building.

    2. _____

3. What is the floor elevation of the existing building. Floor plan size?

    3. _____

4. Office rooms _____ and _____ represent the existing building.

    4. _____

5. Give the width and thickness of the footing at A, B, C on Sheet S-1.

    5. A _____
       B _____
       C _____

6. What size are the footings and piers to be for the columns at F1 and F3, and what reinforcement is required?

    6. _____
        _____
        _____
        _____
        _____
        _____
        _____

7. Give the structural beam size at the east elevation second floor level. In how many sections is this beam?

    7. _____

8. What size column, base plate and stiffeners are to be installed at D on Sheet S-1?

    8. _____
        _____

9. What size beam is called for over the east elevation second floor level entry stair?

    9. _____

10. Give the size of the beam for the north elevation roof framing.

    10. _____
         _____

11. How far above the footing at E on Sheet A-4 does the foundation wall extend before the start of the face brick? Before the concrete block?

    11. _____

12. Is there a concrete foundation wall below the block wall at F on Sheet A-4 or does the block rest on the footing?

    12. _____

13. Give the type of exterior finish material for the four elevations.

13. North _____
East _____
South _____
West _____
_____

14. What width block is used for the wall at G and H on Sheet A-5?

14. G _____
H _____

15. What size lintel is required above the window in room 206? Room 208?

15. 206 _____
_____
208 _____
_____

16. What type windows are called for in the south and east elevations?

16. _____

17. How is the floor for the lower level to be constructed? Second and third levels?

17. _____
_____
_____
_____
_____
_____

18. What materials are used in stair construction?

18. _____
_____
_____

19. How many risers are called for in the stairs between the second and third levels?

19. _____

20. Describe the roof structure and material.

20. _____
_____
_____
_____
_____

21. Give the specifications for the entry door to the offices.

21. _____
_____
_____

22. What type of door is specified for the equipment room on the lower level?

22. _____
_____

23. Describe the finish in room 105.

23. _____
_____
_____

Advance Blueprint Reading Activity 24—2
READING THE PLUMBING, ELECTRICAL AND AIR CONDITIONING
BLUEPRINTS FOR THE OFFICE BUILDING

Refer to Prints 24-1a through 24-1q in the Large Prints Folder to answer the following questions.

1. Give the material used and size of:
   a. Building drain
   b. Building sewer to manhole

   c. Sewer from manhole to public sewer
   d. Public sewer

1. a. _____
   b. _____
      _____
   c. _____
   d. _____

2. How many cleanouts are required for the building drain?

2. _____

3. How many floor drains are required?

3. _____

4. What is the purpose of the sump pump? From where is the water drained into the settling basin?

4. _____
   _____
   _____

5. Give the size of the sump pump basin.

5. _____

6. Toilets for which offices are served by the isometric diagram J on Sheet P-2?

6. _____

7. What is the size of the new water main into the building?

7. _____

8. What type of cold water piping is required under the lower floor, and how is it to be laid?

8. _____

9. By what means is hot water supplied to each lavatory? Give the specifications of the heater.

9. _____
   _____
   _____
   _____
   _____
   _____

10. A fire hydrant is to be added. What is the plumbing contractor's responsibility in connection with this?

10. _____
    _____

Electrical
11. Does the electrical service entrance come in overhead or underground?

11. _____

12. Where does the electrical service enter the building?

12. _____

13. What size conduit and conductors are to be used for the service entrance?

13. _____

14. Give the meter bank specifications.

14. _____
_____
_____
_____

15. How is the meter bank assembly to be grounded?

15. _____
_____
_____

16. Interpret the information at: K, L, M.

16. K _____

L _____

M _____
_____

17. The furnace in office 207 is connected to panel _____ and circuit _____.

17. _____

18. Interpret information at N on Sheet E-3.

18. _____
_____

19. Give the specification for the light in the toilet room of office suite 303.

19. _____
_____
_____

20. What is the specification for the furnace room light in office suite 206?

20. _____
_____

Air Conditioning

21. Give the following for the furnace in office suite 208:
   a. Manufacturer
   b. Model #
   c. Cubic feet/min.
   d. Voltage

21. a. _____
   b. _____
   c. _____
   d. _____

22. Give the following for the cooling condensing unit for office suite 208:
   a. Unit No.
   b. Cooling Cap.
   c. Full Load Amps.
   d. Voltage

22. a. _____
   b. _____
   c. _____
   d. _____

23. List the following for the air conditioning unit
    serving office suite 302:
    a. Location
    b. Manufacturer
    c. Model #
    d. Type Heat
    e. Type Cooling

23. a. _____
    b. _____
    c. _____
    d. _____
    e. _____

24. What size gas supply pipe is brought into the
    building?

24. _____

25. What size gas pipe is run to the roof top to
    serve the furnaces?

25. _____

26. Give the following for office suite 206:
    a. Duct sizes
    b. Specification for supply diffuser

    c. Amount of air at each diffuser.
    d. Size of return air grille

26. a. _____
    b. _____
       _____
       _____
       _____
    c. _____
    d. _____

27. List the specifications for the toilet exhaust
    fans.

27. _____
    _____

28. What size exhaust ducts are required for the
    toilet exhaust fans?

28. _____

29. Give the size of the supply duct and the return
    air duct for the third floor roof top units.

29. Supply _____
    Return _____

30. Where is the thermostat located for office suite
    103-4?

30. _____

Advanced Blueprint Reading Activity 24−3
ESTIMATING THE CONSTRUCTION COSTS FOR THE
OFFICE BUILDING

Refer to Prints 24-1a through 24-1q in the Large Prints Folder to esimate the costs of the complete construction project using the detailed method of estimating. You should include all costs except for the lot which you may assume the owner now owns.

Figure labor and materials for the entire job or figure only those jobs you will directly perform or supervise and get cost estimates on those phases you plan to subcontract. Whichever plan your instructor directs you to follow, you are responsible for the estimates of costs for the entire project.

The following outline will help you in organizing the estimate.

Outline for Preparing a Detailed Estimate

1. Prepare a cost estimate using the form "Estimate Take Off and Cost Sheet" shown in Fig. 24-2. Extra copies of the form appear in the Reference Section. This form was discussed in Unit 20.

2. Consider each of the following items and enter your estimate on the cost estimate form.
    a. Plans and specs
    b. Building permits and fees
    c. Utilities
    d. Excavation and fill
    e. Conc footings, foundations, floors
    f. Unit masonry
    g. Steel beams and columns
    h. Lumber, framing
    i. Lumber, finish
    j. Hardware
    k. Roofing
    l. Windows and screens
    m. Doors and screens
    n. Electric wiring
    o. Electric fixtures
    p. Sheet metal ducts and gutters
    q. Air conditioning
    r. Plumbing
    s. Insulation
    t. Gypsum board
    u. Cabinets
    v. Counter tops
    w. Tile - ceramic, vinyl
    x. Floor sanding
    y. Painting and decorating
    z. Glazing and mirrors
    aa. Floor coverings
    ab. Window cleaning
    ac. Built-ins and accessories
    ad. Range-oven
    ae. Vent hood
    af. Dishwasher
    ag. Garbage disposal
    ah. Tub-shower enclosure
    ai. Clean up
    aj. Insurance protection
    In addition to these items, you should consider business overhead, finance charges and profit on the job.

3. When you have completed your detailed estimate of costs, check your estimate by using the approximate method. Your instructor will assist you in arriving at an approximate cost per unit for your area.

ESTIMATE TAKE OFF AND COST SHEET

Project:_____

Item No.	Item Identification	Location on Job	Cost			Total Cost
			Labor	Material	Equipment	

Fig. 24-2. Form suitable for estimating the cost of construction of the Office Building.

PART 6
REFERENCE SECTION
Dictionary of Terms

ADHESIVE: A cement, glue or other material used to hold two or more parts together.

ADMIXTURES: Materials added to concrete or mortar to alter it in some way as an accelerator, retarder, water-repellent or color.

AGGREGATE: Sand, gravel, rock or material used along with cement to make concrete.

ANCHOR: A device, generally made of metal, used to fasten plates, joists, trusses and other building parts to concrete or masonry.

ANCHOR BOLT: A metal bolt with a threaded end, while the other end usually has an "L" bend. It is embedded in concrete or mortar and is used to hold structural members in place.

ANODIZE: An electrolytic means of coating aluminum or magnesium by oxidizing.

APRON: A piece of trim below the window stool and wall support to conceal edge of wall material. Also: a concrete ramp immediately in front of the garage door.

AREAWAY: The open space around foundation walls, doorways or windows to permit light and air to reach the below-ground-level floors.

ASBESTOS: A mineral material with long thread-like fibers. It is used for exterior wall siding and for fireproofing.

ASHLAR: A stone cut by sawing to a rectangular shape.

ASPHALT: A mineral pitch used for waterproofing roofs and foundation walls. It also is used with crushed rock to pave drives and parking areas.

BACKFILL: To replace earth that has been excavated during construction.

BALUSTERS: Vertical stair members used to support a hand rail.

BALUSTRADE: A row of balusters supporting a common rail.

BATTEN: A narrow strip of wood placed across the joint between two boards, such as siding.

BATTER BOARD: A temporary framework of stakes and horizontal members used in laying out a foundation.

BEAM: A major horizontal structural member used between posts, columns or walls.

BEARING PARTITION: An interior wall which transmits a load from above to a wall, columns or footings below.

BENCH MARK: A point of known elevation, such as a mark cut on a permanent stone or bronze plate set in concrete from which measurements are taken.

BEVEL: A cut on the edge of a board other than 90 degrees.

BEVEL SIDING: A siding material which is tapered from a thick edge to a thinner edge.

BOLSTER: A bent wire device used in holding reinforcing bars in place during the pouring of concrete.

BOND: The holding or gripping force between reinforcing steel and concrete.

BOND BEAM: A steel reinforced concrete masonry beam running horizontally around a masonry wall to provide added strength. Vertical bond beams are formed by inserting reinforcing bars in a cell after the wall is laid and filling with grout.

BONDING: The process of joining two surfaces together, such as with an adhesive.

BRICK VENEER: A brick wall of single brick, usually covering a frame structure.

BRIDGING: The bracing of joists by crossing diagonal pairs of braces.

BUILDING CODE: Laws or regulations set up by building departments of cities, counties, states and Federal Government for uniformity in construction, design and building practices.

CAMBER: A slight vertical curve (arch) formed in a beam or girder to counteract deflection due to loading.

CANT STRIP: A wooden strip used to raise the first course of shingles in plane; an angular board placed at the junction of the roof deck and wall to relieve the sharp angle when the roofing material is installed.

CANTILEVER: A projecting structural member or slab supported at one end only.

CAULK: A nonhardening substance used to seal and waterproof cracks and joints.

CHAIR: A bent wire device used in holding reinforcing bars in place during the pouring of concrete.

CHAMFER: A beveled outside corner or edge on a beam or column.

CHORD: The top and bottom members of open-web joists, and the principal members of trusses as opposed to the diagonals.

CIRCUIT: The electrical path from the source through the components and back to the source.

CIRCUIT BREAKER: A protective device for opening and closing an electric circuit. It opens automatically in case of an overload on the circuit.

CLEANOUT: An opening in the waste pipe of the plumbing system for rodding out the drain. Also: an opening at lower part of the fireplace for removing ashes.

CLERESTORY: A windowed area between roof planes or rising above lower story, to admit light and/or ventilation.

COLD JOINT: Construction joint in concrete occurring at a place where the continuous pouring has been interrupted.

COLLAR BEAM: A tie between two opposite rafters, well above the wall plate.

COLUMN: A vertical structural member.

CONDUCTOR: A material, usually wires, carrying electrical current.

CONDUIT: Metal or fiber pipe used to carry electrical conductors. Also: a metal raceway.

CONSTRUCTION JOINT: Separation between two placements of concrete; a means for keying two sections together.

CONVECTOR: A heat transfer device (radiator) used in a hydronic (hot water) system.

COPING: The top course or cap on a masonry wall protecting the masonry below from water penetration.

CORBEL: A stone, masonry or wood bracket projecting out from a wall.

CORNICE: That part of the roof extending horizontally out from the wall.

COURSE: A horizontal layer of masonry units.

CURTAIN WALL: A non-load bearing wall between columns.

DEAD LOAD: The load on a structure resulting from its own weight of materials and any other fixed loads, such as a roof mounted air conditioner.

DIFFUSER: A grille or register over the air duct opening into a room which controls and directs the flow of air.

DIVERTER: A piece, usually metal, used to direct moisture to a desired path or location.

DORMER: A projection built out from a sloping roof, including one or more vertical windows.

DOWEL: Straight metal bars used to connect or position two sections of concrete or masonry.

DRY WALL: A type of wall covering (gypsum board) used in place of plaster.

DUCT: A round or rectangular pipe, usually metal, used for transferring conditioned air in a heating and cooling system.

EAVES: That portion of the roof which overhangs the wall.

ELEVATION: In surveying, the height of a survey marker above sea level; a measurement on a plot or foundation referenced to a known point. In architectural drafting: the drawing of the front, sides and rear view of a structure.

EXCAVATION: The recess or pit formed by removing the earth in preparation for footings or other foundations.

EXPANSION JOINT: Formed in concrete or masonry units by a bituminous fiber strip to allow for expansion and contraction in materials caused by temperature changes and by shrinkage in materials.

EXTRUSION: Metal which has been shaped by forcing it in the hot or cold state through dies of the desired shape.

FACE BRICK: A select brick especially prepared of clays and chemicals, and fired to produce a desired color and effect for use in the face of a wall.

FASCIA: A finish board nailed to the ends of rafters or lookouts.

FIRE BRICK: A refractory ceramic type brick made to resist high temperatures.

FIRE STOP: A block placed between studs of a wall to prevent a draft and the spread of fire.

FLANGES: The parallel faces of a structural beam joined by the web of the beam.

FLASHING: Sheet metal or other thin material used to prevent moisture from entering a structure, such as around a chimney or wall projecting through the roof.

FLUE: The passageway in a chimney which provides for the escape of smoke, gases and fumes.

FLUSH DOOR: A smooth surface door without panels or molding.

FOOTING: The base of a foundation wall or column which is wider to provide a bearing surface.

FRAMING: The wood or metal structure of a building which gives it shape and strength.

FROST LINE: The depth below the surface subject to freezing.

FURRING: Narrow wood strips fastened to a wall or ceiling for use in nailing finished material.

GABLE: The portion of a wall of a building above the eave line and between the slopes of a double-sloped roof.

GALVANIZED: A coating of zinc on metal to protect it from atmospheric corrosion.

GAUGE: A scale of measurement for wire sizes and sheet metal thickness.

GIRDER: A principal beam supporting other beams.

GLAZING: Installing glass in window sash and doors.

GRADE: The level of the ground around a building; may be designated as existing or finish.

GRADE BEAM: A low foundation wall or a beam, usually at ground level, which provides support for the walls of a building.

GRAVEL STOP: The metal strip at the edge of a built-up roof.

GROUT: A cementitious mixture of high water content, prepared to pour easily into spaces in a masonry wall. Made from portland cement, lime and aggregate, it is used to secure anchor bolts and vertical reinforcing rods in masonry walls.

GUSSET: The piece of metal or plywood used to reinforce a joint of a truss.

HAUNCH: Portion of a beam that increases in depth towards the support.

HEADER: The horizontal structural members over a door or window opening. Also: the joists at the end of an opening in the floor supporting tail joists.

HEADER COURSE: A course of brick laid flat so their long dimension is across the thickness of the wall, and the heads of the course of bricks show on the face of the wall.

HOLLOW CORE DOOR: A lightweight flush door with an interior core of glued strips forming a honeycomb and two exterior smooth panels.

HONEYCOMB: Voids or open spaces left in concrete due to a loss or a shortage of mortar.

HOSE BIBB: A threaded water faucet suitable for fastening a garden hose.

INCREASER: A pipe coupling used between pipes of different sizes.

INSULATING GLASS: A window or door glass consisting of two sheets of glass separated by a sealed air space to reduce heat transfer.

INTEGRALLY-CAST: Element (such as concrete joist and top slab) cast in one piece. (See MONOLITHIC.)

INVERT: The lowest part of the inside of a horizontal pipe.

ISOMETRIC PROJECTION: A pictorial drawing positioned so that its principal axes make equal angles with the plane of projection. Used for some detail and schematic layouts on construction drawings.

JAMB: The top and sides of a door or window frame.

JOIST: One of a series of wood or metal framing members used to support a floor or ceiling.

KALAMEIN DOOR: A metal-covered fireproof door.

KEY: Slotted joint in concrete such as a groove in footing where foundation wall is to be poured.

KILN DRIED: The process of drying wood to a desired moisture content by means of artifically controlled heat and humidity.

KILO: The metric system prefix meaning one thousand times as much as. A kilometer, for example, is 1000 meters.

KILOMETER: One thousand meters. (Equals 5/8 mile.)

KILOWATT: One thousand watts of electricity.

LALLY COLUMN: A vertical steel pipe, usually filled with concrete, used to support beams and girders.

LAMINATED PLASTIC: Layers of cloth or other fiber impregnated with plastic, formed into sheets of the desired thickness and shape with heat and/or pressure. Commonly used for cabinetwork.

LAMINATION: A method of constructing by placing layer upon layer of material and bonding with an adhesive. As a plastic laminate or a wood structural beam.

LANDING: A platform in a flight of stairs to change the direction or to break a long run.

LATH: Gypsum board or metal mesh attached to studs or joists as a base for plaster.

LEDGER: A horizontal strip of wood attached to the side of girders or beams to support joists.

LIFT SLAB: Concrete floor construction in which slabs are cast directly on one another. Each slab is lifted into final position by jack on top of columns. Floors are secured at each floor level by column brackets or collars.

LINTEL: Support for a masonry opening, usually steel angles or special forms.

LIVE LOAD: All movable objects, including persons and equipment in a building.

LONGITUDINAL: With the long dimension of an object or structure.

LOOKOUT: The structural member running from the outside wall to the ends of rafters to carry the plancier or soffit.

LOUVER: A ventilated opening in the attic, usually at gabled end, made of inclined horizontal slats to permit air to pass but to exclude moisture. Also used in doors to ventilate.

MIL: Unit of measuring thickness, 1/1000 of an inch (.001 in.).

MODULAR MEASUREMENT: The design of a structure to use standard size building materials. In the customary system of measurement the module is 4 inches. In the metric system, the recommended module is 100 millimetres.

MONOLITHIC: Concrete cast in one continuous pour.

NEAT CEMENT: A pure cement mixture with no sand or other material added.

NOMINAL SIZE: A general classification term used to designate size of commercial products, such as a 2 x 4. This is not an actual size.

OPEN WEB STEEL JOIST: A truss type joist with top and bottom chords and a web formed of diagonal members. Some manufacturers make a joist with chords of wood and a steel web and refer to it as a truss joist.

ORTHOGRAPHIC PROJECTION: A projection on a picture plane formed by perpendicular projectors from the object to the picture plane. The basis of architectural plan and elevation drawings.

PANEL DOOR: A door of solid frame strips with inset panels.

PARAPET WALL: A wall extending above the roof line.

PARTITION: An interior wall which subdivides an area.

PIER: A heavy column of masonry between two openings used to support other structural members.

232

Dictionary of Terms

PILASTER: A column projecting on the outside or inside of a masonry wall to add strength or decorative effect.

PITCH: Roof slope expressed as a ratio: rise divided by span. Also: spacing between rivets on welds in steel construction.

PLANCIER: The board or panel forming the underside of the eave or cornice.

PLAT: A drawing of a parcel of land indicating lot number, location, boundaries and dimensions. Contains information as to easements and restrictions.

PLENUM: A chamber in an air conditioning system which receives air under pressure before distribution to various ducts.

PLUMB: Perpendicular or vertical. Also: to make a structure vertical.

POCKET DOOR: A door which slides into a partition or wall.

PRESTRESSED CONCRETE: Concrete in which the steel is tensioned (stretched) and anchored to compress the concrete.

PURLIN: A horizontal roof member between the plate and ridge board used to support rafters or trusses.

RACEWAY: Work channels set in floor to receive electrical wiring.

RAFTER: One of a series of structural members of a roof designed to support roof loads. The rafters of a flat roof are sometimes called roof joists.

RECEPTACLE: An electrical outlet to which an appliance or other electrical device may be connected by means of a plug.

REGISTER: A grille used to cover an air duct opening into a room.

REGLET: A long narrow slot in concrete to receive flashing or to serve as anchorage.

ROOF JACK: The sheet metal device placed around a pipe projecting through the roof to prevent moisture from entering.

SADDLE: A small, gable type roof constructed between a vertical surface such as the back of a chimney and a sloped roof to prevent water from standing.

SCALE: A measuring device with graduations for laying off distances. Also: the ratio of size that a structure is drawn, such as: 1/4'' = 1'-0'' which is 1/48 size.

SCHEDULE: A list of details or sizes for building components, such as doors, windows or beams.

SCREED: A template to guide finishers in leveling off the top of fresh concrete. Screeding is ''rough leveling.''

SCUPPER: An opening in the wall at the roof or floor level, permitting water to drain.

SCUTTLE: An opening in the ceiling, providing access to the attic.

SEPTIC TANK: A concrete or metal tank used to reduce raw sewage by bacterial action. Used where no municipal sewage system exists.

SETBACK: The distance from the property boundaries to the building location, required by zoning.

SHEATHING: The boards or panels which cover the studs and rafters of a building before applying the finished siding or roofing material.

SHELF ANGLES: Structural angles which are bolted to a concrete wall to support brick work, stone or terra cotta.

SILL COCK: A valve or water faucet on the outside of a building at about sill height. A hose bibb.

SKEWED: At an angle other than 90 degrees.

SLAB: A flat area of concrete such as a floor or drive.

SLEEPER: Wood strips laid over or embedded in a concrete floor for attaching a finished floor.

SOFFIT: In framing, the underside of a staircase or roof cornice. In masonry, the underside of a beam, lintel or arch.

SOLID CORE DOOR: A flush door having an interior core of solid wood blocks glued together and an exterior of finished veneer paneling or other material, such as hardboard.

SPAN: The distance between supports for joists, beams, girders and trusses.

SPANDREL BEAM: The beam in an exterior wall of a structure.

SPANDREL WALL: The portion of a wall above the head of a window and below the sill of the window above.

SPECS: Short for ''Specifications.'' The written (usually typed) directions issued by architects or engineers to establish general conditions, standards, and detailed instructions which are used with the blueprints.

SQUARE: A unit of measure referring to 100 square feet. Roofing materials and some siding materials are sold on this basis.

STORY: The space between two floors of a building or between a floor and the ceiling above.

STUCCO: A plaster type material consisting of portland cement, sand and water. Used for exterior wall surfaces.

STUD: The vertical framing members of the walls of a structure.

SUPERSTRUCTURE: Frame of the building, usually above grade.

T-BEAM: Beam which has a T-shaped cross section.

TERMITE SHIELD: Sheet metal placed in or on a foundation wall to prevent termites from entering the structure.

THERMOSTAT: An automatic device controlling the operation of an air conditioning system.

TRANSVERSE: Across the short dimension of an object or structure.

TRUSS: A structural unit having such members as beams, bars and ties, usually arranged as triangles. Used for spanning wide spaces.

TYPICAL (TYP): This term, when associated with any dimension or feature, means the dimension or feature applies to the locations that appear to be identical in size and shape unless otherwise noted.

VALLEY: The internal angle formed by two roof slopes coming together.

VENEER: A thin layer of cabinet wood bonded to a plywood or particle board backing.

VENEERED WALL: A single thickness (one wythe) masonry unit wall with a backup wall of frame or other masonry; tied but not bonded to the backup wall.

VENT: A pipe, usually extending through the roof, providing a flow of air to and from the drainage system.

VITRIFIED CLAY TILE: A ceramic tile fired at a high temperature to make it very hard and waterproof.

VOID: Vacant space between material, such as a space in a column.

WEEP HOLES: Small holes in a wall to permit water to exit from behind, as in a retaining wall.

WELDED WIRE FABRIC: Wire mesh fabricated by means of welding the crossing joints.

WORKING DRAWINGS: A set of drawings which provide the necessary details and dimensions to construct the object. May also include the specifications.

WYTHE: A continuous vertical section of masonry, one unit in thickness; sometimes called ''withe'' or ''tier.'' Also: a partition between flues of a chimney.

ACKNOWLEDGMENTS

The author wishes to express his appreciation to Mr. Charles O. Biggs, A.I.A. Architect, for willingly giving of his time and counsel during the organizing and writing of this text. In addition to his professional assistance, Mr. Biggs furnished many of the blueprints contained in the text.

A number of architectural and construction companies also supplied blueprints and technical manuals on all aspects of construction. Appreciation is expressed to the following firms for their assistance: Batson & Associates Architects, Covington, Kentucky; Brick Institute of America, McLean, Virginia; Cypress Speciality Steel Company, Phoenix, Arizona; Environmental Design Consultants: Kral, Zepf, Frietag & Associates, Cincinnati, Ohio; Evans International Homes, Minneapolis, Minnesota; Garlinghouse Plan Service, Topeka, Kansas; Gosnell Development Corporation, Phoenix, Arizona; Robert Ehmet Hayes and Associates, Architects, Fort Mitchell, Kentucky; Helgeson and Biggs, Architects, Inc., Phoenix, Arizona; Henkel, Hovel and Schaefer, Architects-Engineer, Covington, Kentucky; Marathon Steel Company, Phoenix, Arizona; John J. Ross, A.I.A. Architect, Pittsburgh, Pennsylvania; Schweizer Associates Architects, Inc., Environmental Design, Winter Park, Florida; Don Singer, Architect, Broward County, Florida; Smith and Neubek, Architects, Palos Heights, Illinois.

WALTER C. BROWN

Abbreviations

A

A	Air
AB	Anchor Bolt
ABS	Acrylonitrile Butadiene Styrene
AC	Air Conditioning Unit
AC	Alternating Current
ACI	American Concrete Institute
ACOUS	Acoustical
ACP	Asbestos Cement Pipe
AFF	Above Finished Floor
AH	Air Handling Unit
AIA	American Institute of Architects
AISC	American Institute of Steel Construction
AISI	American Iron and Steel Institute
ALT	Alternate
ALUM	Aluminum
AMP	Ampere
AMT	Amount
AP	Access Panel
APPROX	Approximate
ASA	American Standards Association
ASHVE	American Society of Heating and Ventilating Engineers
ASPH	Asphalt
ASTM	American Society of Testing Materials
AUTO.	Automatic
AV	Air Vent
@	At

B

B	Bathroom
B	Bottom
BASMT	Basement
BBL	Barrel(s)
BBR	Base Board Radiation
BD	Board
BEV	Beveled
BF	Bottom of Footing
BLDG	Building
BLK	Block
BLKG	Blocking
BLR	Boiler
BM	Beam
BM	Bench Mark
BOT	Bottom
BP	Base Plate
BP	Blueprint

BR	Bedroom
BR	Brass
BRG	Bearing
BRK	Brick
BRKT	Bracket
BRZ	Bronze
BTU	British Thermal Unit
BUS	Busway
BUZ	Buzzer

C

C	Celsius
C or COND	Conduit
C to C	Center to Center
CAB	Cabinet
CB	Catch Basin
CD	Ceiling Diffuser
CEM	Cement
CER	Ceramic
CFM	Cubic Feet per Minute
CHAM	Chamfer
CHR	Chilled Water Return
CHS	Chilled Water Supply
CI	Cast Iron
CIP	Cast in Place Concrete
CIR	Circuit
CIR BKR	Circuit Breaker
CJ	Construction Joint
CKT	Circuit
CL	Center Line
CL	Closet
CLG	Ceiling
CLK	Caulk
CLR	Clear
cm	Centimetre(s)
CMU	Concrete Masonry Unit
COL	Column
COM	Common
COMP	Composition
CONC	Concrete
CONN	Connection
CONT	Continuous
CONTR	Contractor
CP	Candle Power
CP	Concrete Pipe
CR	Ceiling Register
CSK	Countersink
CT	Ceramic Tile
CU	Copper
CU	Cubic
CW	Cold Water
CWS	Cold Water Supply

D

D	Dryer

DC	Direct Current
DEG	Degree
DET	Detail
DF	Drinking Fountain
DH	Double Hung
DIA or Ø	Diameter
DIAG	Diagram
DIM	Dimension
DIM	Dimmer
DISC	Disconnect
DIV	Division
DMPR	Damper
DN	Down
DO	Ditto
DP	Duplicate
DPG	Damproofing
DR	Dining Room
DR	Drain
DS	Down Spout
DW	Dish Washer
DW	Dry Wall
DWG	Drawing
DWL	Dowel

E

EA	Each
EF	Exhaust Fan
EL	Elevation
ELEC	Electric
ELEV	Elevation
ELVR	Elevator
EMER	Emergency
EMT	Electric Metallic Tubing
ENAM	Enamel
ENC	Enclosure
ENGR	Engineer
ENT	Entrance
EQUIP	Equipment
EST	Estimate
EW	Each Way
EWC	Electric Water Cooler
EXC	Excavate
EXCL	Exclude
EXH	Exhaust
EXIST.	Existing
EXP	Expansion
EXP JT	Expansion Joint
EXT	Exterior
EXTN	Extension

F

F	Fahrenheit
F	Filter
F & I	Furnish and Install
F BRK	Fire Brick

234

Abbreviations

F EXT	Fire Extinguisher	INT	Interior	MTG	Mounting
FAB	Fabricate	INV	Invert	MULL.	Mullion
FAM RM	Family Room				
FAO	Finish All Over		**J**		**N**
FD	Floor Drain				
FDN	Foundation	J	Junction	NEC	National Electric Code
FDR	Feeder	JB	Junction Box	NEMA	National Electrical
FIG.	Figure	JST	Joist		Manufacturers
FIN.	Finish	JT	Joint		Association
FIN. FL	Finish Floor			NF	Near Face
FIX.	Fixture		**K**	NIC	Not in Contract
FL	Flashing			NO.	Number
FL	Floor	K	Kip (1000)	NOM	Nominal
FLEX.	Flexible	K	Kitchen	NOR.	Normal
FLG	Flange	K	Thousand	NTS	Not to Scale
FLG	Flooring	KAL	Kalamein		
FLUOR	Fluorescent	km	Kilometer		**O**
FP	Fireplace	KP	Kick Plate		
FR	Frame	KVA	Kilo Volt Amperes	OBSC GL	Obscure Glass
FRG	Furring	KW	Kilowatt	O. C.	On Center
FS	Floor sink			OD	Outside Diameter
FT	Feet			OF.	Outside Face
FTG	Fitting		**L**	OFF.	Office
FTG	Footing			OPNG	Opening
FURN	Furnish	L	Left	OPP	Opposite
FWH	Frostproof Wall Hydrant	L & PP	Light and Power Panel	OUT.	Outlet
FX WDW	Fixed Window	L CL	Linen Closet	OVHD	Overhead
		LAB	Laboratory		
	G	LAD.	Ladder		**P**
		LAM.	Laminated		
G	Gas	LAT	Lateral	P	Pump
GA	Gauge	LAU	Laundry	PAN.	Pantry
GALV	Galvanized	LAV	Lavatory	PAR.	Parallel
GAR	Garage	LB	Light Beam	PART.	Partition
GB	Grade Beam	LB	Pound	PASS.	Passage
GI	Galvanized Iron	LBR	Lumber	PAV	Paving
GL	Glass	LDG	Landing	PB	Push Button
GL	Glazed	LEV	Level	PC	Pull Chain
GND	Ground	LG	Long	PCS	Pieces
GR	Grade	LPG	Liquefied Petroleum Gas	PERIM	Perimeter
GRAN	Granular	LR	Living Room	PERP	Perpendicular
GRTG	Grating	LT	Light	PH	Phase
GYP	Gypsum	LTH	Lath	PL	Pilot Light
GYP BD	Gypsum Board	LVR	Louver	PL	Plate
				PL	Property Line
	H		**M**	PLAS	Plaster
				PLAT.	Platform
H	Hall	M	Meter	PLBG.	Plumbing
HB	Hose Bibb	M	Thousand	PL GL	Plate Glass
HC	Heating Coil	MAGN	Magnesium	PL HT	Plate Height
HDW	Hardware	MAS	Masonry	PLWD	Plywood
HDWD	Hardwood	MATL	Material	PNL	Panel
HEX.	Hexagonal	MAV	Manual Air Vent	PORC	Porcelain
HGT	Height	MAX.	Maximum	PR	Pair
HM	Hollow Metal	MB	Mixing Boxes (or Units)	PREFAB	Prefabricated
HOR	Horizontal	MC	Medicine Cabinet	PRESS.	Pressure
HP	Horsepower	MCM	Thousand Circular Mills	PRI	Primary
HR	Hour	MDP	Main Distribution Panel	PROP.	Property
HTR	Heater	MECH	Mechanical	PROP.	Proposed
HU	Humidifier	MED	Medium	PRV	Pressure Reducing
HV	Heating and Ventilating	MET.	Metal		(Regulating) Valve
	Unit	MEZZ	Mezzanine	PSI	Pounds per Square Inch
HVAC	Heating and Ventilating	MFR	Manufacturer	PTN	Partition
	and Air Conditioning	MH	Man Hole	PVC	Polyvinyl Chloride
HW	Hot Water	MIN	Minimum	PWR	Power
HWR	Heating Water Return	MISC	Miscellaneous		
HZ	Hertz	ML	Metal Lath		**Q**
		MLDG	Molding		
	I	mm	millimeter	QT	Quarry Tile
		MN	Main		
ID	Inside Diameter	MO	Masonry Opening		**R**
IF	Inside Face	MO	Motor Operated		
INCAND	Incandescent	MOD	Modular	R	Radius
INCL	Include	MOR	Mortar	R	Risers
INCR	Increaser	MS	Manual Starter	RA	Return Air
INSUL	Insulation	MTD	Mounted	RAD	Radiator

RAIC	Royal Architectural Institute of Canada
RCP	Reinforced Concrete Pipe
RD	Roof Drain
RD	Round
REC	Recessed
RECP	Receptacle
REF	Reference
REF	Refrigerator
REG	Register
REINF	Reinforcing
RET	Return
RETG	Retaining
REV	Revision
RFG	Roofing
RGH	Rough
RGH OPNG	Rough Opening
RH	Right Hand
RIO	Rough-In Opening
RM	Room
RV	Relief Valve
RWD	Redwood

S

S	Soil
S	Stretcher
S or SW	Switch
SA	Supply Air
SAD.	Supply Air Diffuser
SAN	Sanitary
SC	Self Closing
SC	Sill Cock
SC	Solid Core
SCH	Schedule
SCR	Screen
SCR	Screw
SCUP	Scupper
SDG	Siding
SDL	Saddle
SEC	Secondary
SECT.	Section
SEL	Select
SER	Service
SEW	Sewer
SG	Supply Air Grille
SH	Sheet
SH	Shower
SH & RD	Shelf and Rod
SHLP	Shiplap
SHTHG	Sheathing
SJ	Steel Joist
SK	Sink
SL	Slate
SL	Sleeve
SM	Sheet Metal
SOV	Shutoff Valve
SP	Soil Pipe
SP	Static Pressure
SP	Sump Pump

SPAN.	Spandrel
SPEC	Specification
SPKR	Speaker
SPR	Sprinkler
SQ	Square
SR	Supply Air Register
SST	Stainless Steel
ST	Storm
STD	Standard
STIFF.	Stiffener
STIR.	Stirrup
STL	Steel
STM	Steam
STN	Stone
STOR	Storage
STR	Straight
STR	Strainer
STRUCT	Structural
SUR	Surface
SUSP	Suspended
SV	Safety Valve
SWG	Standard Wire Gauge
SYM	Symbol

T

T	Thermostat
T & B	Top and Bottom
T & G	Tongue and Groove
TAN.	Tangent
TB	Top of Beam
TC	Terra Cotta
TC	Top of Concrete
TC	Top of Curb
TEL	Telephone
TEMP	Temperature
TEMP	Temporary
TEMPL	Template
TERM.	Terminal
TERR	Terrazzo
TF	Top of Footing
THK	Thick or Thickness
THR	Threaded
THRU.	Through
TJ	Top of Joist
TM	Top of Masonry
TMB	Telephone Mounting Board
TOIL.	Toilet
TP	Top of Pier
TR	Tread
TRANS	Transformer
TS	Tensile Strength
TS	Time Switch
TS	Top of Slab
TV	Television
TV	Turning Vanes
TW	Top of Wall
TYP	Typical

U

UG	Underground
UH	Unit Heater
UL	Underwriters Laboratory
UNFIN	Unfinished
UR	Urinal

V

V	Vent
V	Volts
V-1	Control Valve (as numbered)
VAC	Vacuum
VAR	Varies
VC	Vitrified Clay
VERT	Vertical
VEST.	Vestibule
VOL	Volume
VP	Vitreous Pipe
VTR	Vent Through Roof

W

W	Washer (laundry)
W	Waste
W	Watts
W or WTH	Width
W/	With
W/O	Without
WC	Water Closet
WD	Wood
WDW	Window
WF	Water Fountain
W GL	Wire Glass
WH	Water Heater
WH	Weep Hole
WHR	Watt Hour
WI	Wrought Iron
WM	Water Meter
WP	Waterproofing
WP	Weatherproof
WP	Work Point
WR	Washroom
WR BD	Weather Resistant Board
WS	Weather Strip
WT	Weight
WV	Water Valve
WWF	Welded Wire Fabric

Y

Y	Wye

Z

Z	Zinc

Reference Section

CHARTS AND TABLES

SYMBOLS USED ON CONSTRUCTION DRAWINGS

Charts and tables of symbols commonly used on construction drawings appear on the following pages of this text:

ASTM STANDARD REINFORCING BARS

BAR SIZE DESIGNATION	AREA* Sq. Inches	WEIGHT Pounds per Ft.	DIAMETER* Inches
3	.11	.376	.375
4	.20	.668	.500
5	.31	1.043	.625
6	.44	1.502	.750
7	.60	2.044	.875
8	.79	2.670	1.000
9	1.00	3.400	1.128
10	1.27	4.303	1.270
11	1.56	5.313	1.410
14	2.25	7.650	1.693
18	4.00	13.600	2.257

Current ASTM Specifications provide requirements for sizes 14 and 18 in Grade 60 only.

*Nominal dimensions.

STANDARD TYPES AND SIZES OF WIRE BAR SUPPORTS

SYMBOL	BAR SUPPORT ILLUSTRATION	TYPE OF SUPPORT	STANDARD SIZES
SB	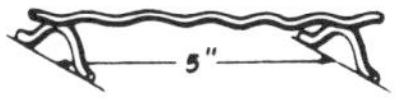	Slab Bolster	¾, 1, 1½, and 2 inch heights in 5 ft. and 10 ft. lengths
SBU*	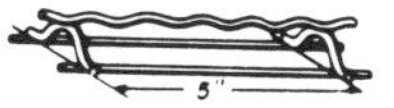	Slab Bolster Upper	Same as SB
BB	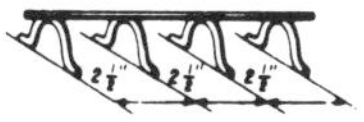	Beam Bolster	1, 1½, 2; over 2" to 5" heights in increments of ¼" in lengths of 5 ft.
BBU*	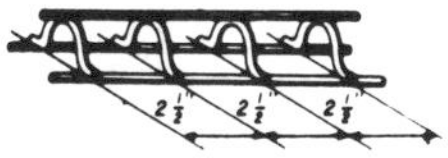	Beam Bolster Upper	Same as BB
BC		Individual Bar Chair	¾, 1, 1½, and 1¾" heights
JC		Joist Chair	4, 5, and 6 inch widths and ¾, 1, and 1½ inch heights
HC		Individual High Chair	2 to 15 inch heights in increments of ¼ in.
HCM*	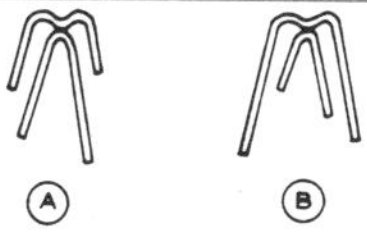	High Chair for Metal Deck	2 to 15 inch heights in increments of ¼ in.
CHC	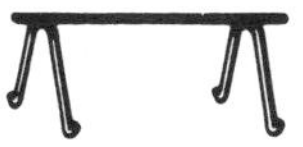	Continuous High Chair	Same as HC in 5 foot and 10 foot lengths
CHCU*	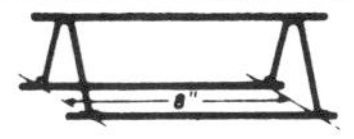	Continuous High Chair Upper	Same as CHC
CHCM*	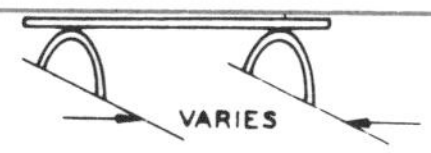	Continuous High Chair for Metal Deck	Up to 5 inch heights in increments of ¼ in.
JCU**	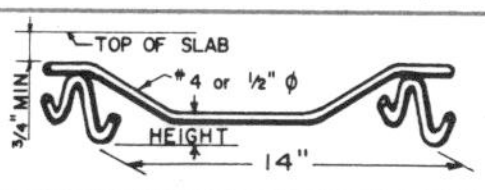	Joist Chair Upper	14" Span. Heights —1" through +3½" vary in ¼" increments

*Available in Class A only, except on special order.
**Available in Class A only, with upturned or end bearing legs.

(Concrete Reinforcing Steel Institute)

METRIC – INCH EQUIVALENTS

INCHES Fractions	INCHES Decimals	MILLIMETERS	INCHES Fractions	INCHES Decimals	MILLIMETERS
	.00394	.1	15/32	.46875	11.9063
	.00787	.2		.47244	12.00
	.01181	.3	31/64	.484375	12.3031
1/64	.015625	.3969	1/2	.5000	12.70
	.01575	.4		.51181	13.00
	.01969	.5	33/64	.515625	13.0969
	.02362	.6	17/32	.53125	13.4938
	.02756	.7	35/64	.546875	13.8907
1/32	.03125	.7938		.55118	14.00
	.0315	.8	9/16	.5625	14.2875
	.03543	.9	37/64	.578125	14.6844
	.03937	1.00		.59055	15.00
3/64	.046875	1.1906	19/32	.59375	15.0813
1/16	.0625	1.5875	39/64	.609375	15.4782
5/64	.078125	1.9844	5/8	.625	15.875
	.07874	2.00		.62992	16.00
3/32	.09375	2.3813	41/64	.640625	16.2719
7/64	.109375	2.7781	21/32	.65625	16.6688
	.11811	3.00		.66929	17.00
1/8	.125	3.175	43/64	.671875	17.0657
9/64	.140625	3.5719	11/16	.6875	17.4625
5/32	.15625	3.9688	45/64	.703125	17.8594
	.15748	4.00		.70866	18.00
11/64	.171875	4.3656	23/32	.71875	18.2563
3/16	.1875	4.7625	47/64	.734375	18.6532
	.19685	5.00		.74803	19.00
13/64	.203125	5.1594	3/4	.7500	19.05
7/32	.21875	5.5563	49/64	.765625	19.4469
15/64	.234375	5.9531	25/32	.78125	19.8438
	.23622	6.00		.7874	20.00
1/4	.2500	6.35	51/64	.796875	20.2407
17/64	.265625	6.7469	13/16	.8125	20.6375
	.27559	7.00		.82677	21.00
9/32	.28125	7.1438	53/64	.828125	21.0344
19/64	.296875	7.5406	27/32	.84375	21.4313
5/16	.3125	7.9375	55/64	.859375	21.8282
	.31496	8.00		.86614	22.00
21/64	.328125	8.3344	7/8	.875	22.225
11/32	.34375	8.7313	57/64	.890625	22.6219
	.35433	9.00		.90551	23.00
23/64	.359375	9.1281	29/32	.90625	23.0188
3/8	.375	9.525	59/64	.921875	23.4157
25/64	.390625	9.9219	15/16	.9375	23.8125
	.3937	10.00		.94488	24.00
13/32	.40625	10.3188	61/64	.953125	24.2094
27/64	.421875	10.7156	31/32	.96875	24.6063
	.43307	11.00		.98425	25.00
7/16	.4375	11.1125	63/64	.984375	25.0032
29/64	.453125	11.5094	1	1.0000	25.4001

METRIC MODULES FOR CONSTRUCTION

The International Standards Organization (ISO) and the National Forest Products Association recommend 100 millimeters as the basic module for the construction industry. This is about the same as the 4 inch module used in the customary system of measurement. The illustration below shows the metric modules and the customary measurement equivalent.

METRIC — FEET CONVERSIONS*										
FEET	1	2	3	4	5	6	7	8	9	10
METRIC MODULE	300 mm 30 cm	600 mm 60 cm 0.6 m	900 mm 90 cm 0.9 m	1200 mm 120 cm 1.2 m	1500 mm 150 cm 1.5 m	1800 mm 180 cm 1.8 m	2100 mm 210 cm 2.1 m	2400 mm 240 cm 2.4 m	2700 mm 270 cm 2.7 m	3000 mm 300 cm 3.0 m
FEET	20	30	40	50	60	70	80	90	100	200
METRIC MODULE	6000 mm 600 cm 6 m	9000 mm 900 cm 9 m	12 000 mm 1200 cm 12 m	15 000 mm 1500 cm 15 m	18 000 mm 1800 cm 18 m	21 000 mm 2100 cm 21 m	24 000 mm 2400 cm 24 m	27 000 mm 2700 cm 27 m	30 000 mm 3000 cm 30 m	60 000 mm 6000 cm 60 m

* RECOMMENDED MODULAR CONVERSIONS

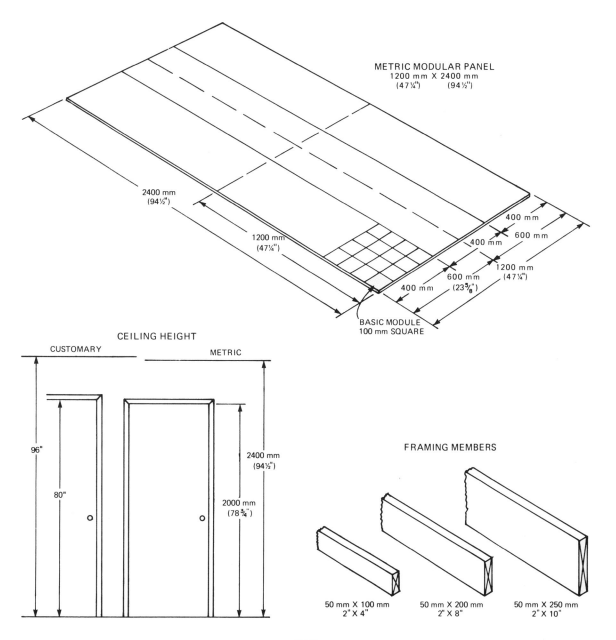

METRIC MODULAR PANEL
1200 mm X 2400 mm
(47¼")   (94½")

2400 mm (94½")

1200 mm (47¼")

400 mm

600 mm

400 mm

1200 mm (47¼")

600 mm (23⅝")

400 mm

BASIC MODULE
100 mm SQUARE

CEILING HEIGHT

CUSTOMARY   METRIC

96"

80"

2400 mm (94½")

2000 mm (78¾")

FRAMING MEMBERS

50 mm X 100 mm
2" X 4"

50 mm X 200 mm
2" X 8"

50 mm X 250 mm
2" X 10"

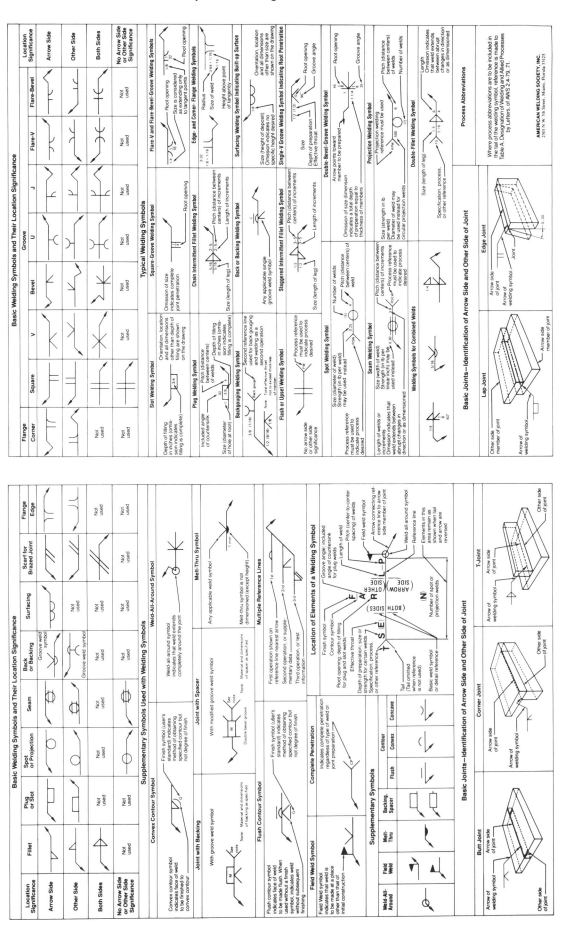

STANDARD WELDING SYMBOLS

ESTIMATE TAKE OFF AND COST SHEET

Project:_____

Item No.	Item Identification	Location on Job	Cost			Total Cost
			Labor	Material	Equipment	

Supplemental Estimate Take Off and Cost Sheet for Part 5.

ESTIMATE TAKE OFF AND COST SHEET

Project:_____

Item No.	Item Identification	Location on Job	Cost			Total Cost
			Labor	Material	Equipment	

Supplemental Estimate Take Off and Cost Sheet for Part 5.

CSI Specifications

BIDDING REQUIREMENTS, CONTRACT FORMS, AND CONDITIONS OF THE CONTRACT

00010 PRE-BID INFORMATION
00100 INSTRUCTIONS TO BIDDERS
00200 INFORMATION AVAILABLE TO BIDDERS
00300 BID FORMS
00400 SUPPLEMENTS TO BID FORMS
00500 AGREEMENT FORMS
00600 BONDS AND CERTIFICATES
00700 GENERAL CONDITIONS
00800 SUPPLEMENTARY CONDITIONS
00900 ADDENDA

Note: The items listed above are not specification sections and are referred to as "Documents" rather than "Sections" in the Master List of Section Titles, Numbers, and Broadscope Section Explanations.

SPECIFICATIONS

DIVISION 1 -- GENERAL REQUIREMENTS

01010 SUMMARY OF WORK
01020 ALLOWANCES
01025 MEASUREMENT AND PAYMENT
01030 ALTERNATES/ALTERNATIVES
01035 MODIFICATION PROCEDURES
01040 COORDINATION
01050 FIELD ENGINEERING
01060 REGULATORY REQUIREMENTS
01070 IDENTIFICATION SYSTEMS
01090 REFERENCES
01100 SPECIAL PROJECT PROCEDURES
01200 PROJECT MEETINGS
01300 SUBMITTALS
01400 QUALITY CONTROL
01500 CONSTRUCTION FACILITIES AND TEMPORARY CONTROLS
01600 MATERIAL AND EQUIPMENT
01650 FACILITY STARTUP/COMMISSIONING
01700 CONTRACT CLOSEOUT
01800 MAINTENANCE

DIVISION 2 -- SITEWORK

02010 SUBSURFACE INVESTIGATION
02050 DEMOLITION
02100 SITE PREPARATION
02140 DEWATERING
02150 SHORING AND UNDERPINNING
02160 EXCAVATION SUPPORT SYSTEMS
02170 COFFERDAMS
02200 EARTHWORK
02300 TUNNELING
02350 PILES AND CAISSONS
02450 RAILROAD WORK
02480 MARINE WORK
02500 PAVING AND SURFACING
02600 UTILITY PIPING MATERIALS
02660 WATER DISTRIBUTION
02680 FUEL AND STEAM DISTRIBUTION
02700 SEWERAGE AND DRAINAGE
02760 RESTORATION OF UNDERGROUND PIPE
02770 PONDS AND RESERVOIRS
02780 POWER AND COMMUNICATIONS
02800 SITE IMPROVEMENTS
02900 LANDSCAPING

DIVISION 3 -- CONCRETE

03100 CONCRETE FORMWORK
03200 CONCRETE REINFORCEMENT
03250 CONCRETE ACCESSORIES
03300 CAST-IN-PLACE CONCRETE
03370 CONCRETE CURING
03400 PRECAST CONCRETE
03500 CEMENTITIOUS DECKS AND TOPPINGS
03600 GROUT
03700 CONCRETE RESTORATION AND CLEANING
03800 MASS CONCRETE

DIVISION 4 -- MASONRY

04100 MORTAR AND MASONRY GROUT
04150 MASONRY ACCESSORIES
04200 UNIT MASONRY
04400 STONE
04500 MASONRY RESTORATION AND CLEANING
04550 REFRACTORIES
04600 CORROSION RESISTANT MASONRY
04700 SIMULATED MASONRY

DIVISION 5 -- METALS

05010 METAL MATERIALS
05030 METAL COATINGS
05050 METAL FASTENING
05100 STRUCTURAL METAL FRAMING
05200 METAL JOISTS
05300 METAL DECKING
05400 COLD FORMED METAL FRAMING
05500 METAL FABRICATIONS
05580 SHEET METAL FABRICATIONS
05700 ORNAMENTAL METAL
05800 EXPANSION CONTROL
05900 HYDRAULIC STRUCTURES

DIVISION 6 -- WOOD AND PLASTICS

06050 FASTENERS AND ADHESIVES
06100 ROUGH CARPENTRY
06130 HEAVY TIMBER CONSTRUCTION
06150 WOOD AND METAL SYSTEMS
06170 PREFABRICATED STRUCTURAL WOOD
06200 FINISH CARPENTRY
06300 WOOD TREATMENT
06400 ARCHITECTURAL WOODWORK
06500 STRUCTURAL PLASTICS
06600 PLASTIC FABRICATIONS
06650 SOLID POLYMER FABRICATIONS

DIVISION 7 -- THERMAL AND MOISTURE PROTECTION

07100 WATERPROOFING
07150 DAMPPROOFING
07180 WATER REPELLENTS
07190 VAPOR RETARDERS
07195 AIR BARRIERS
07200 INSULATION
07240 EXTERIOR INSULATION AND FINISH SYSTEMS
07250 FIREPROOFING
07270 FIRESTOPPING
07300 SHINGLES AND ROOFING TILES
07400 MANUFACTURED ROOFING AND SIDING
07480 EXTERIOR WALL ASSEMBLIES
07500 MEMBRANE ROOFING
07570 TRAFFIC COATINGS
07600 FLASHING AND SHEET METAL
07700 ROOF SPECIALTIES AND ACCESSORIES
07800 SKYLIGHTS
07900 JOINT SEALERS

DIVISION 8 -- DOORS AND WINDOWS

08100 METAL DOORS AND FRAMES
08200 WOOD AND PLASTIC DOORS
08250 DOOR OPENING ASSEMBLIES
08300 SPECIAL DOORS
08400 ENTRANCES AND STOREFRONTS
08500 METAL WINDOWS
08600 WOOD AND PLASTIC WINDOWS
08650 SPECIAL WINDOWS
08700 HARDWARE
08800 GLAZING
08900 GLAZED CURTAIN WALLS

DIVISION 9 -- FINISHES

09100 METAL SUPPORT SYSTEMS
09200 LATH AND PLASTER
09250 GYPSUM BOARD
09300 TILE
09400 TERRAZZO
09450 STONE FACING
09500 ACOUSTICAL TREATMENT
09540 SPECIAL WALL SURFACES
09545 SPECIAL CEILING SURFACES
09550 WOOD FLOORING
09600 STONE FLOORING
09630 UNIT MASONRY FLOORING
09650 RESILIENT FLOORING
09680 CARPET
09700 SPECIAL FLOORING
09780 FLOOR TREATMENT
09800 SPECIAL COATINGS
09900 PAINTING
09950 WALL COVERINGS

DIVISION 10 -- SPECIALTIES

10100 VISUAL DISPLAY BOARDS
10150 COMPARTMENTS AND CUBICLES
10200 LOUVERS AND VENTS
10240 GRILLES AND SCREENS‾
10250 SERVICE WALL SYSTEMS
10260 WALL AND CORNER GUARDS
10270 ACCESS FLOORING
10290 PEST CONTROL
10300 FIREPLACES AND STOVES
10340 MANUFACTURED EXTERIOR SPECIALTIES
10350 FLAGPOLES
10400 IDENTIFYING DEVICES
10450 PEDESTRIAN CONTROL DEVICES
10500 LOCKERS
10520 FIRE PROTECTION SPECIALTIES
10530 PROTECTIVE COVERS
10550 POSTAL SPECIALTIES
10600 PARTITIONS
10650 OPERABLE PARTITIONS
10670 STORAGE SHELVING
10700 EXTERIOR PROTECTION DEVICES FOR OPENINGS
10750 TELEPHONE SPECIALTIES
10800 TOILET AND BATH ACCESSORIES
10880 SCALES
10900 WARDROBE AND CLOSET SPECIALTIES

DIVISION 11 -- EQUIPMENT

11010 MAINTENANCE EQUIPMENT
11020 SECURITY AND VAULT EQUIPMENT
11030 TELLER AND SERVICE EQUIPMENT
11040 ECCLESIASTICAL EQUIPMENT
11050 LIBRARY EQUIPMENT
11060 THEATER AND STAGE EQUIPMENT
11070 INSTRUMENTAL EQUIPMENT
11080 REGISTRATION EQUIPMENT
11090 CHECKROOM EQUIPMENT
11100 MERCANTILE EQUIPMENT
11110 COMMERCIAL LAUNDRY AND DRY CLEANING EQUIPMENT
11120 VENDING EQUIPMENT
11130 AUDIO-VISUAL EQUIPMENT
11140 VEHICLE SERVICE EQUIPMENT
11150 PARKING CONTROL EQUIPMENT
11160 LOADING DOCK EQUIPMENT
11170 SOLID WASTE HANDLING EQUIPMENT
11190 DETENTION EQUIPMENT
11200 WATER SUPPLY AND TREATMENT EQUIPMENT
11280 HYDRAULIC GATES AND VALVES
11300 FLUID WASTE TREATMENT AND DISPOSAL EQUIPMENT
11400 FOOD SERVICE EQUIPMENT
11450 RESIDENTIAL EQUIPMENT
11460 UNIT KITCHENS
11470 DARKROOM EQUIPMENT
11480 ATHLETIC, RECREATIONAL, AND THERAPEUTIC EQUIPMENT
11500 INDUSTRIAL AND PROCESS EQUIPMENT
11600 LABORATORY EQUIPMENT
11650 PLANETARIUM EQUIPMENT
11660 OBSERVATORY EQUIPMENT
11680 OFFICE EQUIPMENT
11700 MEDICAL EQUIPMENT
11780 MORTUARY EQUIPMENT
11850 NAVIGATION EQUIPMENT
11870 AGRICULTURAL EQUIPMENT

DIVISION 12 -- FURNISHINGS

12050 FABRICS
12100 ARTWORK
12300 MANUFACTURED CASEWORK
12500 WINDOW TREATMENT
12600 FURNITURE AND ACCESSORIES
12670 RUGS AND MATS
12700 MULTIPLE SEATING
12800 INTERIOR PLANTS AND PLANTERS

DIVISION 13 -- SPECIAL CONSTRUCTION

13010 AIR SUPPORTED STRUCTURES
13020 INTEGRATED ASSEMBLIES
13030 SPECIAL PURPOSE ROOMS
13080 SOUND, VIBRATION, AND SEISMIC CONTROL
13090 RADIATION PROTECTION
13100 NUCLEAR REACTORS
13120 PRE-ENGINEERED STRUCTURES
13150 AQUATIC FACILITIES
13175 ICE RINKS
13180 SITE CONSTRUCTED INCINERATORS
13185 KENNELS AND ANIMAL SHELTERS
13200 LIQUID AND GAS STORAGE TANKS
13220 FILTER UNDERDRAINS AND MEDIA
13230 DIGESTER COVERS AND APPURTENANCES
13240 OXYGENATION SYSTEMS
13260 SLUDGE CONDITIONING SYSTEMS
13300 UTILITY CONTROL SYSTEMS
13400 INDUSTRIAL AND PROCESS CONTROL SYSTEMS
13500 RECORDING INSTRUMENTATION
13550 TRANSPORTATION CONTROL INSTRUMENTATION
13600 SOLAR ENERGY SYSTEMS
13700 WIND ENERGY SYSTEMS
13750 COGENERATION SYSTEMS
13800 BUILDING AUTOMATION SYSTEMS
13900 FIRE SUPPRESSION AND SUPERVISORY SYSTEMS
13950 SPECIAL SECURITY CONSTRUCTION

DIVISION 14 -- CONVEYING SYSTEMS

14100 DUMBWAITERS
14200 ELEVATORS
14300 ESCALATORS AND MOVING WALKS
14400 LIFTS
14500 MATERIAL HANDLING SYSTEMS
14600 HOISTS AND CRANES
14700 TURNTABLES
14800 SCAFFOLDING
14900 TRANSPORTATION SYSTEMS

DIVISION 15 --MECHANICAL

15050 BASIC MECHANICAL MATERIALS AND METHODS
15250 MECHANICAL INSULATION
15300 FIRE PROTECTION
15400 PLUMBING
15500 HEATING, VENTILATING, AND AIR CONDITIONING
15550 HEAT GENERATION
15650 REFRIGERATION
15750 HEAT TRANSFER
15850 AIR HANDLING
15880 AIR DISTRIBUTION
15950 CONTROLS
15990 TESTING, ADJUSTING, AND BALANCING

DIVISION 16 -- ELECTRICAL

16050 BASIC ELECTRICAL MATERIALS AND METHODS
16200 POWER GENERATION - BUILT-UP SYSTEMS
16300 MEDIUM VOLTAGE DISTRIBUTION
16400 SERVICE AND DISTRIBUTION
16500 LIGHTING
16600 SPECIAL SYSTEMS
16700 COMMUNICATIONS
16850 ELECTRIC RESISTANCE HEATING
16900 CONTROLS
16950 TESTING

INDEX